THE FRONTIERS COLLECTION

THE FRONTIERS COLLECTION

The books in this collection are devoted to challenging and open problems at the forefront of modern science, including related philosophical debates. In contrast to typical research monographs, however, they strive to present their topics in a manner accessible also to scientifically literate non-specialists wishing to gain insight into the deeper implications and fascinating questions involved. Taken as a whole, the series reflects the need for a fundamental and interdisciplinary approach to modern science. Furthermore, it is intended to encourage active scientists in all areas to ponder over important and perhaps controversial issues beyond their own speciality. Extending from quantum physics and relativity to entropy, consciousness and complex systems—the Frontiers Collection will inspire readers to push back the frontiers of their own knowledge.

More information about this series at http://www.springer.com/series/5342

For a full list of published titles, please see back of book or springer.com/series/5342

Anthony Aguirre · Brendan Foster
Zeeya Merali
Editors

IT FROM BIT OR BIT FROM IT?

On Physics and Information

FQXi
Foundational Questions Institute

Springer

Editors
Anthony Aguirre
Department of Physics
University of California
Santa Cruz, CA
USA

Zeeya Merali
Foundational Questions Institute
New York, NY
USA

Brendan Foster
Foundational Questions Institute
New York, NY
USA

ISSN 1612-3018 ISSN 2197-6619 (electronic)
THE FRONTIERS COLLECTION
ISBN 978-3-319-36075-1 ISBN 978-3-319-12946-4 (eBook)
DOI 10.1007/978-3-319-12946-4

Springer Cham Heidelberg New York Dordrecht London

Softcover reprint of the hardcover 1st edition 2015

Printed on acid-free paper

Springer International Publishing AG Switzerland is part of Springer Science+Business Media (www.springer.com)

Preface

This book is a collaborative project between Springer and The Foundational Questions Institute (FQXi). In keeping with both the tradition of Springer's Frontiers Collection and the mission of FQXi, it provides stimulating insights into a frontier area of science, while remaining accessible enough to benefit a non-specialist audience.

FQXi is an independent, nonprofit organization that was founded in 2006. It aims to catalyze, support, and disseminate research on questions at the foundations of physics and cosmology.

The central aim of FQXi is to fund and inspire research and innovation that is integral to a deep understanding of reality, but which may not be readily supported by conventional funding sources. Historically, physics and cosmology have offered a scientific framework for comprehending the core of reality. Many giants of modern science—such as Einstein, Bohr, Schrödinger, and Heisenberg—were also passionately concerned with, and inspired by, deep philosophical nuances of the novel notions of reality they were exploring. Yet, such questions are often overlooked by traditional funding agencies.

Often, grant-making and research organizations institutionalize a pragmatic approach, primarily funding incremental investigations that use known methods and familiar conceptual frameworks, rather than the uncertain and often interdisciplinary methods required to develop and comprehend prospective revolutions in physics and cosmology. As a result, even eminent scientists can struggle to secure funding for some of the questions they find most engaging, while younger thinkers find little support, freedom, or career possibilities unless they hew to such strictures.

FQXi views foundational questions not as pointless speculation or misguided effort, but as critical and essential inquiry of relevance to us all. The Institute is dedicated to redressing these shortcomings by creating a vibrant, worldwide community of scientists, top thinkers, and outreach specialists who tackle deep questions in physics, cosmology, and related fields. FQXi is also committed to engaging with the public and communicating the implications of this foundational research for the growth of human understanding.

As part of this endeavor, FQXi organizes an annual essay contest, which is open to everyone, from professional researchers to members of the public. These contests are designed to focus minds and efforts on deep questions that could have a profound impact across multiple disciplines. The contest is judged by an expert panel and up to 20 prizes are awarded. Each year, the contest features well over a hundred entries, stimulating ongoing online discussion for many months after the close of the contest.

We are delighted to share this collection, inspired by the 2013 contest, "It from Bit or Bit from It?" In line with our desire to bring foundational questions to the widest possible audience, the entries, in their original form, were written in a style that was suitable for the general public. In this book, which is aimed at an interdisciplinary scientific audience, the authors have been invited to expand upon their original essays and include technical details and discussion that may enhance their essays for a more professional readership, while remaining accessible to non-specialists in their field.

FQXi would like to thank our contest partners: The Gruber Foundation, The John Templeton Foundation, and Scientific American. The editors are indebted to FQXi's scientific director, Max Tegmark, and managing director, Kavita Rajanna, who were instrumental in the development of the contest. We are also grateful to Angela Lahee at Springer for her guidance and support in driving this project forward.

2014

Anthony Aguirre
Brendan Foster
Zeeya Merali

Contents

Chapter 1
Introduction

Anthony Aguirre, Brendan Foster and Zeeya Merali

> *'It from bit' symbolizes the idea that every item of the physical world has at bottom—at a very deep bottom, in most instances—an immaterial source and explanation; that what we call reality arises in the last analysis from the posing of yes-no questions and the registering of equipment-evoked responses; in short, that all things physical are information-theoretic in origin and this is a participatory universe.*
>
> John Archibald Wheeler (1989)

Over the past century, there has been a steady progression away from thinking about physics, at its deepest level, as a description of material objects and their interactions, and towards physics as a description of the evolution of information about, and in, the physical world. Information theory encompasses the apparently inherent probabilistic nature of quantum mechanics, our statistical understanding of thermodynamical systems, and computer science, where the encoding of data is described classically using rules laid out by Claude Shannon. Recent years have seen an explosion of interest at the nexus of physics and information, driven by the information age in which we live and by developments in quantum information theory and computer science.

The idea that information is more fundamental than the matter that conveys it was famously encapsulated by physicist John Archibald Wheeler in the phrase "It from Bit". Wheeler was profoundly aware of the peculiar relationship between information and the measurements made by observers of quantum systems: He proposed a "delayed-choice experiment" using light—which has since been demonstrated in the laboratory—in which it appears that choices made by an experimenter *after* a

A. Aguirre (✉)
Department of Physics, University of California, Santa Cruz, CA, USA
e-mail: aguirre@scipp.ucsc.edu

B. Foster · Z. Merali
Foundational Questions Institute, New York, NY, USA
e-mail: foster@fqxi.org

Z. Merali
e-mail: merali@fqxi.org

A. Aguirre et al. (eds.), *It From Bit or Bit From It?*,
The Frontiers Collection, DOI 10.1007/978-3-319-12946-4_1

measurement has been made could influence the result of that measurement, although it had been carried out earlier. Wheeler also coined the term "black hole" to describe the cosmic objects that swallow everything with the misfortune to fall close enough, including light. It has since been discovered that black holes appear to have the thermodynamic properties of entropy and temperature. The fate of information about the infalling matter and light, after the black hole has evaporated away, is the source of the so-called black-hole information paradox, which has caused debate among physicists for decades.

But is information truly fundamental? If information forms the most basic layer of reality, what does it refer to? With these issues in mind, in 2013, we posed the following broad question to professional researchers and to the public in FQXi's annual essay contest: "It From Bit or Bit From It?" (The name, it should be noted, was inspired, in part, by the prize-winning entry, "Bit from It", submitted by Julian Barbour to a previous essay contest. Barbour's essay is included in this volume.) We hoped to inspire entrants to tackle the issues of what information is, and to clarify its relation to reality. Other open questions include how nature (the universe and the things therein) store and process information, and how understanding information elucidates physics and vice-versa.

The contest was a resounding success, attracting over 180 entries. Some attempted to answer Wheeler's question directly; others took on the tricky task of trying to unpick just what Wheeler meant with that question: Did he mean that the actual universe literally emerges as an evolving answer to a series of yes/no questions? Or was he trying to express that all physical laws eventually reduce to binary computations? Does the question make any sense at all?

This volume showcases 18 of the winning essays from the 2013 contest and also Barbour's contribution from a previous contest. We begin our selection with first prize winner, Matt Leifer, who directly tackles the connection between information, our formulations of probability theory, and the foundations of quantum theory. In Chap. 2, Leifer argues that a generalization of probability theory is needed to understand how *It* may derive from *Bit*, but that such a formulation is also compatible with the seemingly conflicting notion that *Bit* arises from *It*.

Chapters 3 and 4 cement the link with quantum theory further by modifying Wheeler's pronouncement to state that *It* derives from *Qubit* (a qubit is a quantum version of a binary digit that can take the values of 0 and 1 simultaneously). Giacomo D'Ariano makes the case that particles, spacetime, and more, arise from quantum interactions. Michel Planat explores Wheeler's fascination with the part played by the observer in creating reality by investigating the machinery behind the act of quantum measurements and their information content.

The role of the observer is further explored in Chaps. 5 and 6. Cristinel Stoica discusses the delayed-choice experiment, in particular, and how in a participatory universe, as outlined by Wheeler, the observer may affect the emergence of physical laws. Kevin Knuth re-casts the universe as a network of information-based influences where the laws of physics derive, in part, from the inferences made by observers.

Information theory spans not just quantum theory, but also classical physics, including aspects of thermodynamics and computer science. This feature is exploited by the authors of Chaps. 7–9. Carlo Rovelli argues that Shannon's notion of relative information between two physical systems serves as an objective foundation for both statistical mechanics and quantum mechanics and, by extension, could be at the base of a naturalistic picture of the world. Paul Borrill combines elements of Boltzmann's statistical account of thermodynamics and Shannon's theory to provide insight into the quantum phenomenon of entanglement and, in particular, the ambiguous nature of time in quantum systems. Yutaka Shikano uses similar considerations to question whether a physical system can be completely understood once its Hamiltonian or Langragian (descriptions of the energy of the system) are known.

In Chap. 10, Douglas Singleton, Elias Vagenas and Tao Zhu consider both the quantum and thermodynamic natures of black holes in order to tackle the information paradox directly.

The role of spacetime as a carrier of information is considered in Chaps. 11–13. Torsten Asselmeyer-Maluga proposes a model, in which a smooth spacetime contains a discrete amount of information, that could explain aspects of cosmology, such as the accelerated inflation of the early universe. Craig Hogan describes a radical experiment that could directly test whether reality has finite information content. Sean Gryb considers arguments for whether spacetime is continuous or discrete and then presents a third alternative in which scale is meaningless and only shape is of fundamental interest.

Some entrants argued against Wheeler's stance that *It* derives from *Bit*, and these contributions appear in Chaps. 14–18. Mark Feeley cautions that giving up the notion that material things are fundamental leaves us in a world without material causes. The team of Angelo Bassi, Saikat Ghosh and Tejinder Singh contend that Wheeler's statement is based on the assumption that quantum theory is inherently probabilistic and that this preconception can be challenged. Ken Wharton similarly argues that there may be a realistic framework that underlies quantum theory. Julian Barbour, by analyzing the definition of information, and Ian Durham, through searching for an underlying quantum principle that could guide how reality may emerge from information, each independently favor the contrary statement that turns Wheeler's view on its head: *Bit* from *It*.

Finally, in Chap. 19, William McHarris, proposes a compromise view, based on analysis of the standard interpretation of quantum mechanics in light of chaos theory. He ultimately argues that *It* is founded on *Bit*, which in turns derives from *It*, and so on ...To try to unpack this further is akin, he claims, to asking, which came first, the chicken or the egg?

In summary, this book brings together authors from a range of physics and mathematical disciplines, including those who specialize in the study of quantum foundations, particle physics, astrophysics and cosmology, nuclear physics, physics engineering, and quantum gravity. This diversity reflects the immense reach of information theory across multiple research areas. As a result, each essay offers a unique perspective on the foundational role of information.

Chapter 2
"It from Bit" and the Quantum Probability Rule

M.S. Leifer

Abstract I argue that, on the subjective Bayesian interpretation of probability, "it from bit" requires a generalization of probability theory. This does not get us all the way to the quantum probability rule because an extra constraint, known as noncontextuality, is required. I outline the prospects for a derivation of noncontextuality within this approach and argue that it requires a realist approach to physics, or "bit from it". I then explain why this docs not conflict with "it from bit". This version of the essay includes an addendum responding to the open discussion that occurred on the FQXi website. It is otherwise identical to the version submitted to the contest.

Wheeler's "It from Bit"

> It from bit. Otherwise put, every it—every particle, every field of force, even the spacetime continuum itself—derives its function, its meaning, its very existence entirely—even if in some contexts indirectly—from the apparatus-elicited answers to yes or no questions, binary choices, bits.
>
> It from bit symbolizes the idea that every item of the physical world has at bottom—at a very deep bottom, in most instances—an immaterial source and explanation; that what we call reality arises in the last analysis from the posing of yes-no questions and the registering of equipment-evoked responses; in short, that all things physical are information-theoretic in origin and this is a participatory universe.
>
> — J. A. Wheeler [1]

John Wheeler's "it from bit" is a thesis about the foundations of quantum theory. It says that the things that we usually think of as real—particles, fields and even spacetime—have no existence independent of the questions that we ask about them. When a detector clicks it is not registering something that was there independently of the experiment. Rather, the very act of setting up the detector in a certain way—the choice of question—is responsible for the occurrence of the click. It is only the act of

M.S. Leifer (✉)
Perimeter Institute for Theoretical Physics,
31 Caroline Street, North Waterloo, ON N2L 2Y5, Canada
e-mail: matt@mattleifer.info

A. Aguirre et al. (eds.), *It From Bit or Bit From It?*,
The Frontiers Collection, DOI 10.1007/978-3-319-12946-4_2

asking questions that causes the answers to come into being. This idea is perhaps best illustrated by Wheeler's parable of the game of twenty questions (surprise version).

> You recall how it goes—one of the after-dinner party sent out of the living room, the others agreeing on a word, the one fated to be a questioner returning and starting his questions. "Is it a living object?" "No." "Is it here on earth?" "Yes." So the questions go from respondent to respondent around the room until at length the word emerges: victory if in twenty tries or less; otherwise, defeat.
>
> Then comes the moment when we are fourth to be sent from the room. We are locked out unbelievably long. On finally being readmitted, we find a smile on everyone's face, sign of a joke or a plot. We innocently start our questions. At first the answers come quickly. Then each question begins to take longer in the answering—strange, when the answer itself is only a simple "yes" or "no." At length, feeling hot on the trail, we ask, "Is the word 'cloud'?" "Yes," comes the reply, and everyone bursts out laughing. When we were out of the room, they explain, they had agreed not to agree in advance on any word at all. Each one around the circle could respond "yes" or "no" as he pleased to whatever question we put to him. But however he replied he had to have a word in mind compatible with his own reply—and with all the replies that went before. No wonder some of those decisions between "yes" and "no" proved so hard!
>
> — J. A. Wheeler [2]

Wheeler proposed "it from bit" as a clue to help us answer the question "How come the quantum?", i.e. to derive the mathematical apparatus of quantum theory from a set of clear physical principles. In this essay, I discuss whether "it from bit" implies the quantum probability rule, otherwise known as the Born rule, which would get us part of the way towards answering Wheeler's question.

My main argument is that, on the subjective Bayesian interpretation of probability, "it from bit" requires a generalized probability theory. I explain why this is not ruled out by the common claim that classical probability theory is not to be violated on pain of irrationality.

In the context of quantum theory, "it from bit" does not quite get us all the way to the Born rule because the latter mandates a further constraint known as noncontextuality. The prospects for understanding noncontextuality as a rationality requirement or an empirical addition are slim. Extra physical principles are needed and I argue that these must be about the nature of reality, rather than the nature of our knowledge. This seems to conflict with "it from bit" as it requires and agent-independent reality, suggesting "bit from it". I argue that there is no such conflict because the sense of "it" used in "it from bit" is different from the sense used in "bit from it".

The Interpretation of Probability

Since von Neumann's work on quantum logic and operator algebras [3, 4], it has been known that quantum theory can be viewed as a generalization of probability theory [5, 6]. If we want to understand what this tells us about the nature of reality then we will need to adopt a concrete theory of how probabilities relate to the world, which is the job of an interpretation of probability theory.[1] Three main classes of interpretation have arisen to meet this need: frequentism (probability is long run relative frequency), epistemic probability (probabilities represent the knowledge, information, or beliefs of a decision making agent), and objective chance (probabilities represent a kind of law of nature or a disposition for a system to act in a certain way). Getting into the details of all these options would take us too far afield, but a few comments are in order to explain why adopting my preferred epistemic interpretation, known as subjective Bayesianism,[2] is not a crazy thing to do.

Frequentism is still popular amongst physicists, but it has largely been abandoned by scholars of the philosophy of probability. It is not able to handle single-case probabilities, e.g. the probability that civilization will be destroyed by a nuclear war, and it leads to a bizarre reading of the law of large numbers that does not do the explanatory work required of it.[3] A common position in the philosophy of probability is that subjective Bayesianism is more satisfactory, but that it needs to be backed up by some theory of objective chance in order to account for probabilistic laws.[4] My own view is that subjective Bayesianism suffices on its own, but whether or not one believes in objective chance is irrelevant for the present discussion, since objective chances need to be connected to epistemic probabilities in some way in order to explain how we can come to know statistical laws. The usual way of doing this is via David Lewis' principal principle [13]. One of the implications of this is that objective chances must have the same mathematical structure as subjective Bayesian probabilities. Therefore, if we can argue that "it from bit" requires a modification of subjective Bayesian probability then the same will apply to objective chances as well. It is also worth noting that several modern interpretations of quantum theory adopt subjective Bayesianism, including "Quantum Bayesianism" [14–17] and the Deutsch-Wallace variant of many-worlds [18–21] amongst others [22].

Subjective Bayesian Probability

Subjective Bayesianism says that probabilities represent the degrees of belief of a decision making agent, who is conventionally described in the second person as "you". Degrees of belief are measured by looking at your behaviour, e.g. your willingness to enter into bets. The claim is that if you do not structure your beliefs according to the axioms of probability theory then you are irrational. There are various ways of deriving this, differing in their simplicity and sophistication. For ease of exposition, I base my discussion on the simplest approach, known as the Dutch book argument.

---

1 See [7] for an accessible introduction and [8] for a collection of key papers.

2 Subjective Bayesianism has its origins in [9, 10]. An accessible introduction is [11].

3 See [12] for a critique of frequentism in statistics.

4 This view originates with David Lewis [13].

The Dutch book argument defines your degree of belief in the occurrence of an uncertain event E as the value $\$p(E)$ you consider to be a fair price for a lottery ticket that pays \$1 if E occurs and nothing if it does not. "Fair price" here means that you would be prepared to buy or sell any number of these tickets at that price and that you would be prepared to do this in combination with fair bets on arbitrary sets of other events. Your degrees of belief are said to be irrational if a malicious bookmaker can force you to enter into a system of bets that would cause you to lose money whatever the outcome, despite the fact that you consider them all fair. Otherwise, your degrees of belief are said to be rational. The Dutch book argument then shows that your degrees of belief are rational if, and only if, they satisfy the usual axioms of probability theory. These axioms are:

- **Background framework**: There is a set Ω, called the sample space, containing the most fine-grained events you might be asked to bet on, e.g. if you are betting on the outcome of a dice roll then $\Omega = \{1, 2, 3, 4, 5, 6\}$. In general, an event is a subset of Ω, e.g. the event that the dice roll comes out odd is $\{1, 3, 5\}$. For simplicity, we assume that Ω is finite. The set of events forms a Boolean algebra, which just means that it supports the usual logical notions of AND, OR and NOT.
- **A1**: For all events $E \subseteq \Omega$, $0 \leq p(E) \leq 1$.
- **A2**: For the certain event Ω, $p(\Omega) = 1$.
- **A3**: If $E \cap F = \emptyset$, i.e. E and F cannot both happen together, then $p(E \cup F) = p(E) + p(F)$, where $E \cup F$ means the event that either E or F occurs.

For illustrative purposes, here is the part of the argument showing that violations of **A1** and **A2** are irrational. Consider an event E and suppose contra **A1** that $p(E) < 0$. This means that you would be willing to sell a lottery ticket that pays out on E to the bookie for a negative amount of money, i.e. you would pay her $\$p(E)$ to take the ticket off your hands. Now, if E occurs you will have to pay the bookie \$1 so in total you will have paid her $\$1 + p(E)$, and if E does not occur you will have paid her a total of $\$p(E)$. Either way, you will lose money so having negative degrees of belief is irrational. A similar argument shows that having degrees of belief larger than 1 is irrational. Now suppose contra **A2** that $p(\Omega) < 1$. Then, you would be prepared to sell the lottery ticket for $\$p(\Omega)$ and pay out \$1 if Ω occurs. However, since Ω is certain to occur, you will always end up paying out, which leaves you with a loss.

Is Probability Theory Normative?

Based on this kind of argument, many subjective Bayesians regard probability theory as akin to propositional logic.[5] In logic, you start with a set of premises that you regard as true, and then you use the rules of logic to figure out what other propositions must be true or false as a consequence. If you fail to abide by those truth values then there is an inconsistency in your reasoning. However, there is nothing in logic that

[5] For example, see [23] where this argument is made repeatedly.

tells you what premises you have to start with. The premises are simply the input to the problem and logic tells you what else they compel you to believe. Similarly, subjective probability does not tell you what numerical value you must assign to any uncertain event,[6] but given some of those values as input, it tells you what values you must assign to other events on pain of inconsistency, the inconsistency here being exposed in the form of a sure loss. Like logic, subjective Bayesians regard probability theory as normative rather than descriptive, i.e. they claim that you should structure your degrees of belief about uncertain events according to probability theory if you aspire to be ideally rational, but not that humans actually do structure their beliefs in this way. In fact, much research shows that they do not [25].

The normative view of probability theory presents a problem if we want to view quantum theory as generalized probability because it implies that it is irrational to use anything other than conventional probability theory to reason about uncertain events. Fortunately, the normative view is not just wrong, but obviously wrong. Unlike logic, it is easy to come up with situations in which the Dutch book argument has no normative force. Because of this, the idea that it might happen in quantum theory too is not particularly radical.

For example, the Dutch book argument requires that you view the fair price for selling a lottery ticket to be the same as the fair price for buying it. In reality, people are more reluctant to take on risk than they are to maintain a risk for which they have already paid the cost. Therefore, the fair selling price might be higher than the fair buying price. This leads to the more general theory of upper and lower probabilities wherein degrees of belief are represented by intervals on the real line rather than precise numerical values [26].

At this point, I should address the fact that the Dutch book argument is not the only subjective Bayesian derivation of probability theory, so its defects may not be shared with the other derivations. The most general subjective arguments for probability theory are formulated in the context of decision theory, with Savage's axioms being the most prominent example [27]. These take account of things like the fact that you may be risk averse and your appreciation of money is nonlinear, e.g. $1 is worth more to a homeless person than a billionaire, so they replace the financial considerations of the Dutch book argument with the more general concept of "utility". However, what all these arguments have in common is that they are hedging strategies. They start with some set of uncertain events and then they introduce various decision scenarios that you could be faced with where the consequences depend on uncertain events in some way, e.g. the prizes in a game that depends on dice rolls. Importantly, these arguments only work if the set of decision scenarios is rich enough. They ask you to consider situations in which the prizes for the various outcomes are chopped up and exchanged with each other in various ways. For example, in the Dutch book argument this comes in the form of the idea that you must be prepared to buy or sell arbitrarily many tickets for arbitrary sets of events at the fair price. The arguments then conclude

[6] This is where it differs from objective Bayesianism [24], which asserts that there is a unique rational probability that you ought to assign. However, defining such a unique probability is problematic at best.

that if you do not structure your beliefs according to probability theory then there is some decision scenario in which you would be better off had you done so and none in which you would be worse off. However, in real life, it is rather implausible that you would be faced with such a rich set of decision scenarios. More often, you know something in advance about what decisions you are going to be faced with. This is why decision theoretic arguments are hedging strategies. They start from a situation in which you do not know what decisions you are going to be faced with and then they ask you to consider the worst possible scenario. If you know for sure that this scenario is not going to come up then the arguments have no normative force.

As an example, consider the following scenario. There is a coin that is going to be flipped exactly once. You have in your possession \$1 and you are going to be forced to bet that dollar on whether the coin will come up heads or tails, with a prize of \$2 if you get it right. You do not have the option of not placing a bet. How should you structure your beliefs about whether the coin will come up heads or tails? If the decision theoretic arguments applied then we would be forced to say that you must come up with a precise numerical value for the probability of heads $p(H)$. However, it is clear that the cogitation involved in coming up with this number is completely pointless in this scenario. All you need to know is the answer to a single question. Do you think heads is more likely to come up than tails? Your decision is completely determined by this answer, which is just a single bit of information rather than a precise numerical value.

It should be clear from this that the decision theoretic arguments are not as strongly normative as the laws of logic. Instead they are *conditionally* normative, i.e. normative if the decision scenarios envisaged in the argument are all possible.

"It from Bit" Implies a Generalized Probability Theory

A universe that obeys "it from bit" is a universe in which not all conceivable decision scenarios are possible. To explain this, consider again Wheeler's parable of twenty questions (surprise version) and imagine that you are observing the game passively, placing bets with a bookmaker on the side as it proceeds. To make things more analogous to quantum theory, imagine that the respondents exit the room as soon as they have answered their question, never to be heard from again. We might imagine that they are sent through a wormhole into a region of spacetime that will forever be outside of our observable universe and that the wormhole promptly closes as soon as they enter it. This rules out the possibility that we might ask them about what they would have answered if they had been asked a different question, since in quantum theory we generally cannot find out what the outcome of a measurement that we did not actually make would have been.

Suppose that, at some point in the game, you make a bet with the bookie that the object that the fifth respondent has in mind is a dove. However, what actually happens is that the questioner asks "Is it white?" and the answer comes back "yes", whereupon the fifth respondent is whisked off to the far corners of the universe.

Now, although the answer "yes" is consistent with the object being a dove, this is not enough to resolve the bet as there are plenty of other conceivable white objects. As in Wheeler's story, suppose that the last question asked is "Is it a cloud?" and that the answer comes back "yes". In the usual version of twenty questions this would be enough to resolve the bet in the bookie's favor because all the respondents are thinking of a common object. However, in the surprise version this is not the case. It could well be that "dove" was consistent with all the answers given so far at the time we made the bet, and that the fifth respondent was actually thinking of a dove. We can never know and so the bet can never be resolved. It has to be called off and you should get a refund.

Whilst the bet described above is unresolvable, other bets are still jointly resolvable, e.g. a bet on whether the fifth respondent was thinking of a white object together with a bet on whether the last respondent was thinking of a cloud. The set of bets that is jointly resolvable depends on the sequence of questions that is actually asked by the questioner. If you want to develop a hedging strategy ahead of time, then you need to consider all possible sequences of questions that might be asked to ensure that you cannot be forced into a sure loss for any of them.

For the subjective Bayesian, the main lesson of this is that, in general, only certain subsets of all possible bets are jointly resolvable. Define a *betting context* to be a set of events such that bets on all of them are jointly resolvable and to which no other event can be added without violating this condition. It is safe to assume that each betting context is a Boolean algebra, since, if we can find out whether E occurred at the same time as finding out whether F occurred, then we can also determine whether they both occurred, whether either one of them occurred, and whether they failed to occur, so we can define the usual logical notions of AND, OR and NOT. However, unlike in conventional probability theory, there need not be a common algebra on which all of the events that occur in different betting contexts are jointly defined. Because of this, the Dutch book argument has normative force within a betting context, but it does not tell us how probabilities should be related across different contexts. Therefore, our degrees of belief should be represented by a set of probability distributions $p(E|\mathcal{B})$, one for each betting context $\mathcal{B}$.[7]

This framework can be applied to quantum theory where the betting contexts represent sets of measurements that can be performed together at the same time. The details of this are rather technical, so they are relegated to Appendix. The probabilities that result from this are more general than those allowed by quantum theory. To get uniquely down to the Born rule, we need an extra constraint, known as *noncontextuality*. This says that there are certain pairs of events from different betting

[7] Despite the notation, $p(E|\mathcal{B})$ is not a conditional probability distribution because there need not be a common algebra on which all the events are defined. Some authors do not consider this to be a generalization of probability theory [21, 28], since all we are saying is that we have a bunch of probability distributions rather than just one. However, such systems can display nonclassical features such as violations of Bell inequalities and no-cloning [29] so they are worthy of the name "generalization" if anything is.

contexts, $E \in \mathcal{B}$ and $F \in \mathcal{B}'$, that must always be assigned the same probability $p(E|\mathcal{B}) = p(F|\mathcal{B}')$. Therefore, we need to explain how such additional constraints can be understood.

Noncontextuality in Subjective Bayesianism

One option is that noncontextuality could simply be posited as an additional fundamental principle. Previous Dutch book arguments for the Born rule have done essentially this [14, 22]. However, subjective Bayesians do not accept fundamental constraints on probabilities beyond those required by rationality. Imposing such constraints would be like saying that you are allowed to construct a logical argument providing one of your starting premises is "the car is red", but if you start from "the car is yellow" then any argument you make is logically invalid. Additional constraints on probabilities are contingent facts about your state of belief, just as logical premises are contingent facts about the world. Therefore, noncontextuality needs to be derived in some way.

One possibility is that noncontextuality follows from logical equivalence, i.e. if quantum theory always assigns the same probability to E in context $\mathcal{B}$ and F in $\mathcal{B}'$ then these should be regarded as equivalent logical statements, in the same sense that E and NOT (NOTE) are equivalent in a Boolean algebra.[8] Logical equivalence implies that it ought to be possible to construct a Dutch book that results in a sure loss if $p(E|\mathcal{B}) \neq p(F|\mathcal{B}')$. This can only be done if you are willing to accept that the occurrence of E in betting context $\mathcal{B}$ makes it necessary that F would have occurred had the betting context been $\mathcal{B}'$ and vice versa. If this is the case, then you will agree that a bet made on F in betting context $\mathcal{B}'$ should also pay out if the betting context was in fact $\mathcal{B}$ and the event E occurred and vice versa. If this is the case, then the bookie can construct a Dutch book against $p(E|\mathcal{B}) < p(F|\mathcal{B}')$ by buying a ticket from you that pays out on E and selling a ticket that pays out on F. The payouts on these tickets will be the same, so you will lose money in this transaction. By exchanging the roles of E and F, there would be a Dutch book against $p(E|\mathcal{B}) > p(F|\mathcal{B}')$ as well.

This strategy hinges on whether it is reasonable to make counterfactual assertions, i.e. assertions about what would have happened had the betting context been different. However, "it from bit" declares such counterfactuals meaningless because it says that there is no answer to questions that have not been asked. Even if we do not accept "it from bit", the Kochen-Specker theorem [30] implies that counterfactual assertions cannot all respect noncontextuality, i.e. there would have to be pairs of events $E \in \mathcal{B}$ and $F \in \mathcal{B}'$ such that if E occurs in $\mathcal{B}$ then F would not have occurred in $\mathcal{B}'$ even though quantum theory asserts that $p(E|\mathcal{B}) = p(F|\mathcal{B}')$ always holds. We conclude that noncontextuality of probability assignments cannot be a rationality requirement.

---

[8] Pitowsky attempts to argue along these lines [22], unsuccessfully in my view.

Another possibility is that noncontextuality could be adopted simply because we have performed many quantum experiments and have always observed relative frequencies in accord with the Born rule. Although probabilities are not identified with relative frequencies in subjective Bayesianism, it still offers an account of statistical inference wherein observing relative frequencies causes probabilities to be updated. If certain technical conditions hold, probability assignments will converge to the observed relative frequency in the limit of a large number of trials. Therefore, we could assert that noncontextuality is a brute empirical fact.[9]

The problem with this is that it provides no explanation of why noncontextuality holds. If we accept this, we might as well just give up and say that the only reason why we believe any physical theory is because it matches the observed relative frequencies. It would be like saying that the reason why the Maxwell-Boltzmann distribution applies to a box of gas is because we have sampled many molecules from such boxes and always found them to be approximately Maxwell-Boltzmann distributed. This belies the important explanatory role of stationary distributions in equilibrium statistical mechanics, and would be of no help in understanding why nonequilibrium systems tend to equilibrium. Similarly, the Born rule appears to be playing an important structural role in quantum theory that calls for an explanation.

The remaining option is to view noncontextuality as arising from physical, as opposed to logical, equivalence. The Dutch book rationality criterion is usually expressed as the requirement that you should not enter into bets that lead to a sure loss by logical necessity, but it is equally irrational to enter into a bet that you believe will lead to a sure loss, whether or not that belief is a logical necessity. Because of this, the argument that you should assign probability one to the certain event equally applies to events that you only believe to be certain, regardless of whether that belief is correct. Now, belief in the laws of physics entails certainty about statements that follow from the laws so this can be the origin of constraints on probability assignments.

To illustrate, suppose you believe that Newtonian mechanics is true and that there is a single particle system with a given Hamiltonian. This means that you are committed to propositions of the form "If the particle initially occupies phase space point (x_0, p_0) then at time t it occupies the solution to Hamilton's equations $(x(t), p(t))$ with initial condition (x_0, p_0)". If you bet on such propositions at anything more or less than even odds then you believe that you will lose money with certainty. Importantly, this type of argument can also imply constraints on events that you are not certain about. For example, if you assign a phase space region some probability and then compute the endpoints of the trajectories for all points in that region at a later time then the region formed by the end points must be assigned the same probability at that later time. This shows that the need to assign equal probabilities to different events can sometimes be derived from the laws of physics.

Crucially, this sort of argument can only really be made to work if there is an objectively existing external reality. There needs to be some sort of "quantum stuff" such that events that are always assigned the same probability correspond to physically equivalent states of this stuff. In the context of the many-worlds interpretation,

---

[9] This has been suggested in the context of the many-worlds interpretation [28].

the Deutsch-Wallace [18–21] and Zurek [31, 32] derivations of the Born rule are arguments of this type, where the quantum stuff is simply the wavefunction.

"It from Bit" or "Bit from It"?

We have arrived at the conclusion that noncontextuality must be derived in terms of an analysis of the things that objectively exist. This implies a realist view of physics, or in other words "bit from it", which seems to conflict with "it from bit". Fortunately, this conflict is only apparent because "it" is being used in different senses in "it from bit" and "bit from it". The things that Wheeler classifies as "it" are things like particles, fields and spacetime. They are things that appear in the fundamental ontology of classical physics and hence are things that only appear to be real from our perspective as classical agents. He does not mention things like wavefunctions, subquantum particles, or anything of that sort. Thus, there remains the possibility that reality is made of quantum stuff and that the interaction of this stuff with our question asking apparatus, also made of quantum stuff, is what causes the answers (particles, fields, etc.) to come into being. "It from bit" can be maintained in this picture provided the answers depend not only on the state of the system being measured, but also on the state of the stuff that comprises the measuring apparatus. Thus, we would end up with "it from bit from it", where the first "it" refers to classical ontology and the second refers to quantum stuff.

Conclusion

On the subjective Bayesian view, "it from bit" implies that probability theory needs to be generalized, which is in accord with the observation that quantum theory is a generalized probability theory. However, "it from bit" does not get us all the way to the quantum probability rule. A subjective Bayesian analysis of noncontextuality indicates that it can only be derived within a realist approach to physics. At present, this type of derivation has only been carried out in the many-worlds interpretation, but I expect it can be made to work in other realist approaches to quantum theory, including those yet to be discovered.

Addendum

In editing this essay for publication, I wanted to hew as closely as possible to the version submitted to the contest, so I have decided to address the discussion that occurred on the FQXi website in this addendum. I also address some comments made by Kathryn Laskey in private correspondence, because I think she addressed one of the issues particularly eloquently. I am grateful to my colleagues and co-entrants for their thoughtful comments. It would be impossible to address all of them here, so I restrict attention to some of the most important and frequently raised issues. Further details can be found on the FQXi comment thread [33].

Noncontextuality

Both Jochen Szangolies and Ian Durham expressed confusion at my usage of the term "noncontextuality", which derives from Gleason's theorem [34]. Due to the Kochen-Specker theorem [30], it is often said that quantum theory is "contextual", so how can this be reconciled with my claim that the Born rule is "noncontextual"?

In Gleason's theorem, noncontextuality means that the same probability should be assigned to the same projection operator, regardless of the context that it is measured in. Here, by context, I mean the other projection operators that are measured simultaneously. So, as in the example given in the technical endnotes, $|2\rangle$ should receive the same probability regardless of whether it is measured as part of the basis $\{|0\rangle, |1\rangle, |2\rangle\}$ or the basis $\{|+\rangle, |-\rangle, |2\rangle\}$, where $|\pm\rangle = \frac{1}{\sqrt{2}}(|0\rangle \pm |1\rangle)$. From the perspective of this essay, Gleason's theorem says that, for Hilbert spaces of dimension three or larger, the only probability assignments compatible with both the Dutch book constraints and noncontextuality are those given by the Born rule. In this sense the Born rule is "noncontextual" and indeed it is the only viable probability rule that is.

On the other hand, the Kochen-Specker theorem concerns the assignment of definite values to the outcomes of measurements. Instead of assigning probabilities to projectors, the aim is to assign them values 0 or 1 in such a way that, for any set of orthogonal projectors that can occur together in a measurement, exactly one of them gets the value 1. This is to be done noncontextually, which means that whatever value a projector is assigned in one measurement context, it must be assigned the same value in all other contexts in which it occurs. The Kochen-Specker theorem says that this cannot be done.

The two theorems are related because 0 and 1 are examples of probability assignments, albeit extremal ones. As first pointed out by Bell [35], Gleason's theorem actually implies the conclusion of the Kochen-Specker theorem by the following argument. For any quantum state, the Born rule never assigns $0/1$ probabilities to every single projector. Gleason's theorem implies that, in dimension three and higher, the only noncontextual probability assignments are given by the Born rule. Therefore, for these dimensions, there can be no noncontextual probability assignment that only assigns $0/1$ probabilities.

From this it should be apparent that the noncontextuality assumption of the Kochen-Specker theorem is the same as in Gleason's theorem, only that it is specialized to $0/1$ probability assignments. The additional assumption that the probabilities must be either 0 or 1 is called *outcome determinism*, so the Kochen-Specker theorem shows that it is impossible to satisfy both outcome determinism and noncontextuality at the same time (in addition to the Dutch book constraints).

Based on this, people often loosely say that the Kochen-Specker theorem shows that quantum theory is "contextual" and this is the source of the confusion. However, adopting contextual value assignments is only one way of resolving the contradiction entailed by the Kochen-Specker theorem, the other being to drop outcome determinism. It is therefore perfectly consistent to say that the Born rule is noncontextual but that any model that assigns definite values to every observable cannot be.

Scientific Realism

Scientific realism is the view that our best scientific theories should be thought of as describing an objective reality that exists independently of us. My argument ends up endorsing the realist position, as it concludes that the world must be made of some objectively existing "quantum stuff".

There are good a priori reasons for believing in scientific realism that are independent of the specifics of quantum theory, and hence independent of the argument given in this essay (see [36] for a summary and [37] for a more detailed treatment of these arguments). Most people who believe in scientific realism are probably swayed by these arguments rather than anything to do with the details of quantum theory.

As pointed out by Ken Wharton, this would seem to open the possibility of short-circuiting my argument. Why not simply make the case for scientific realism via one or more of the a priori arguments? From this it follows that the world must consist of some objectively existing stuff, and hence "bit from it".

Whilst I agree that this is a valid line of argument, my intention was not to provide an argument that would convince realists, for whom "bit from it" is a truism. It is evident from the popularity of interpretations of quantum theory that draw inspiration from the Copenhagen interpretation, which I collectively call *neo Copenhagen* interpretations, that not everyone shares such strong realist convictions. Wheeler's "it from bit" is usually read as a neo-Copenhagen principle. It says that what we usually call reality derives from the act of making measurements rather than from something that exists independently of us. As I argue in the essay, "it from bit" can be given a more realist spin by interpreting "it" as referring to an emergent, effective classical reality rather than to the stuff that the world is made of at the fundamental level. Nevertheless, most endorsers of "it from bit" are likely to have the neo Copenhagen take on it in mind.

The most effective way of arguing against any opponent is to start from their own premises and show that they lead to the position they intend to oppose. This is much more effective than arguing against their premises on a priori grounds as it is evident from the fact that they have chosen those premises that the opponent does not find such a priori arguments compelling. My aim here is to do this with "it from bit"—a premise accepted by many neo Copenhagenists—and to argue that it needs to be supplemented with realism, or "bit from it", in order to obtain a compelling derivation of the Born rule. This presents a greater challenge to the neo Copenhagen view than simply rehashing the existing arguments for realism. I expect my fellow realists to find this line of argument overly convoluted, but it is not really aimed at them.

Has Probability Theory Really been Generalized?

In this essay, I argued that "it from bit" requires a generalization of probability theory. Specifically, I argued that there are a number of different betting contexts $\mathcal{B}_1, \mathcal{B}_2, \ldots$, that within each betting context the Dutch book argument implies a well defined probability measure over the Boolean algebra of events in that context, but that it does not imply any constraints on events across different betting contexts. This gives rise to a theory in which there are a number of different Boolean algebras, each of which has its own probability measure, instead of there being just one probability measure over a single Boolean algebra. Giacomo Mauro D'Ariano, Howard Barnum and Kathryn Laskey (the latter in private correspondence) questioned whether it is really necessary to think of this as a generalization of probability theory.

D'Ariano's method for preserving probability theory is to assign probabilities to the betting contexts themselves. That is, we can build a sample space of the form $(\mathcal{B}_1 \times \Omega_{\mathcal{B}_1}, \mathcal{B}_2 \times \Omega_{\mathcal{B}_2}, \ldots)$, where $\Omega_{\mathcal{B}_j}$ is the sample space associated with betting context $\mathcal{B}_j$. We can then just specify an ordinary probability measure over this larger space, and the separate probability measures for each context would then be obtained by conditioning on $\mathcal{B}_j$.

I admit that this can always be done formally, but conceptually one might not want to regard betting contexts as the kind of thing that should be assigned probabilities. They are defined by the sequences of questions that we decide to ask, so one might want to regard them as a matter of "free choice". To avoid the thorny issue of free will, we can alternatively imagine that the betting context is determined by an adversary. Recall that, for a subjective Bayesian, assigning a probability to an event means being willing to bet on that event at certain odds. Therefore, assigning probabilities to betting contexts means you should be willing to bet on which context will occur. However, if the bookie is also the person who gets to choose the betting context after all such bets are laid, then she can always do so in such a way as to make your probability assignments to the betting contexts as inaccurate as possible. Therefore, there are at least some circumstances under which it would not be meaningful to assign probabilities to betting contexts.

Laskey's response is quite different. She simply denies that what I have described deserves the name "generalization of probability theory". Since her comments were made in private communication, with her permission I reproduce them here.

> Let me first take issue with your statement that quantum theory requires generalizing probability theory because the Boolean algebras of outcomes are different in different betting contexts. Dependence of the Boolean algebra of outcomes on the betting context is by no means restricted to quantum theory. It happens all the time in classical contexts—in fact, it's a fixture of our daily life. Ever see the movie "Sliding Doors"? The Boolean algebra of

outcomes I face today would be totally different had I not chosen to marry my husband; had I taken a different job when I came out of grad school; had my husband and I not had four children; had I not chosen an academic career; had I not put myself into a position in which other people depend on me to put food on the table; or any number of other might-have-been in my life.

Consider, for example, a town facing the question of whether to zone a given area for residential development or to put a wind farm there. If the town chooses residential development, we might have, for example, a probability distribution over a Boolean algebra of values of the average square footage of homes in the area. There would be no such Boolean algebra if we build the wind farm. (I am specifically considering averages because they are undefined when N is zero.) If we choose the wind farm, we would have a distribution over the average daily number of kilowatt hours of wind-powered electricity generated by the wind farm. There would be no such Boolean algebra if we choose the residential development. What is the intrinsic difference between this situation and the case of a quantum measurement, in which the algebra of post-measurement states depends on the experiment the scientist chooses to conduct?

Just about any time we make a decision, the Boolean algebra of possible future states of the world is different for each choice we might make. Decision theorists are accustomed to this dependence of possible outcomes on the decision. It does not mean we need to generalize probability theory. It simply means we have a different Boolean algebra conditional on some contexts than conditional on others.

In some ways this is a matter of semantics. I argue in the essay that the breakdown of some of the usual conventions of probability theory is commonplace and should not be surprising. We just disagree on whether this deserves the name "generalization".

My argument for a generalization of probability theory is mainly directed against dogmatic Bayesians who endorse the view that ordinary probability theory on a single Boolean algebra is not to be violated on pain of irrationality. There are plenty of dogmatic Bayesians still around. If modern Bayesians have a more relaxed attitude then that is all to the good as far as I am concerned. However, I do think it is worth making the argument specifically in the context of physics, as physicists are often a bit timid about drawing implications for the foundations of probability from their subject, and I do not think they should be if violations of the standard framework are commonplace.

I therefore do not wish to spill too much ink over whether or not the bare-bones theory of multiple Boolean algebras should be called a generalization of probability theory. However, quantum theory has much more structure than this, in the form of Hilbert space structure and the noncontextuality requirement. For me, the more important question is whether quantum theory should be viewed as a generalization of probability theory.

Must Quantum Theory be Viewed as a Generalization of Probability Theory?

The short answer to this is no. The underdetermination of theory by evidence implies that there will always be several ways of formulating a theory that are empirically equivalent. We can always apply D'Ariano's trick or take Laskey's view, since they apply to any set of probabilities on separate Boolean algebras, and quantum theory is just a restriction on that set. Therefore, I cannot argue that it is a logical necessity to view quantum theory as a generalized probability theory, but I can argue that it is more elegant, simpler, productive, etc. to do so.

As an analogy, note that it is also not logically necessary to view special relativity as ruling out the existence of a luminiferous ether. Instead, one can posit that there exists an ether, that it picks out a preferred frame of reference in which it is stationary, but that forces act upon objects in such a way to make it impossible to detect motion relative to the ether, e.g. they cause bodies moving relative to the ether to contract in just such a way as to mimic relativistic length contraction. This theory makes the exact same predictions as special relativity and is often called the Lorentz ether theory.[10] Special relativity is normally regarded as superior to the Lorentz ether theory because the latter seems to require a weird conspiracy of forces in order to protect the existence of an entity that cannot be observed. The former has proved to be a much better guide to the future development of physics. What I want to argue is that not adopting a view in which quantum theory is a generalization of probability theory is analogous to adopting the Lorentz ether theory, i.e. it is consistent but a poor guide to the future progress of physics.

When we add Hilbert spaces and noncontextuality into the mix, Gleason's theorem implies that our beliefs can be represented by a density operator ρ on Hilbert space, at least if the Hilbert space is of dimension three of higher. I have argued elsewhere that regarding the density operator as a true generalization of a classical probability distribution leads to an elegant theory which unifies a lot of otherwise disparate quantum phenomena [6]. Here, I will confine myself to a different argument, based on the quantum notion of entropy.

Classically, the entropy of a probability distribution $\boldsymbol{p} = (p_1, p_2, \ldots, p_n)$ over a finite space is given by

$$H(\boldsymbol{p}) = -\sum_j p_j \log p_j. \tag{2.1}$$

Up to a multiplicative constant, this describes both the Shannon (information theoretic) entropy and the Gibbs (thermodynamic) entropy. In other words, it describes the degree of compressibility of a string of digits drawn from independent instances of the probability distribution $\boldsymbol{p}$ and also quantifies the amount of heat that must be dissipated in a thermodynamic transformation. In quantum theory, the entropy of a density operator is given by the von Neumann entropy,

$$S(\rho) = -\mathrm{Tr}\,(\rho \log \rho)\,, \tag{2.2}$$

which is the natural way of generalizing the classical entropy if you think of density operators as the quantum generalization of probability distributions. It turns out

[10] It is similar to the theory in which Lorentz first derived his eponymous transformations, although, unlike the theory described here, the actual theory proposed by Lorentz failed to agree with special relativity in full detail.

that this plays the same role in quantum theory as the classical entropy does in classical theories, i.e. it is both the information theoretic entropy, quantifying the compressibility of quantum states drawn from a source described by ρ, and it is the thermodynamic entropy, quantifying the heat dissipation in a thermodynamic transformation.

Now, this definition of quantum entropy only really makes sense on the view that density operators are generalized probability measures. What would we get if we took D'Ariano or Laskey's views instead?

On D'Ariano's view we have a well-defined classical probability distribution, just over a larger space that includes the betting contexts. If this is just an ordinary classical probability distribution then arguably we should just use the formula for the classical entropy, although one might want to marginalize over the betting contexts first. This is not the von Neumann entropy, and it does not seem to quantify anything of relevance to quantum information or thermodynamics.

On Laskey's view we just have a bunch of unrelated probability distributions over different betting contexts. Should we take the entropy of just one of these and, if so, which one? None of them seems particularly preferred. Should we take some kind of weighted average of all of them and, if so, what motivates the weighting given that betting contexts are not assigned probabilities? Arguably the only relevant betting context is the one we actually end up in, so one should just apply the classical entropy formula to this context, but this is unlikely to match the von Neumann entropy.[11]

Now admittedly, there is probably some convoluted way of getting to the von Neumann entropy in these other approaches, just as there is a way of understanding the Lorentz transformations in the Lorentz ether theory, but I expect that it would look ad hoc compared to treating density operators as generalized probability distributions.

In summary, I am arguing that quantum theory should be regarded as a bona fide generalization of probability theory, not out of logical necessity, but because doing so gives the right quantum generalizations of classical concepts. People who view things in this way are liable to make more progress in quantum information theory, quantum thermodynamics, and beyond, than those who do not. In this sense I think the situation is analogous to adopting special relativity over Lorentz ether theory.

Technical Endnotes

In general, a betting context $\mathcal{B}$ is a Boolean algebra, which we take to be finite for simplicity. All such algebras are isomorphic to the algebra generated by the subsets of some finite set $\Omega_{\mathcal{B}}$, where AND is represented by set intersection, OR by union, and NOT by complement.

[11] It will match if we are lucky enough to choose the context that minimizes the classical entropy, but again there is no motivation for doing this in Laskey's approach.

In quantum theory, a betting context corresponds to a set of measurements that can be performed together that is as large as possible. A measurement is represented by a self-adjoint operator M and all such operators have a spectral decomposition of the form

$$M = \sum_j \lambda_j \Pi_j, \tag{2.3}$$

with eigenvalues λ_j and orthogonal projection operators Π_j that sum to the identity $\sum_j \Pi_j = I$. The eigenvalues are the possible measurement outcomes and, when the system is assigned the density operator ρ, the Born rule states that the outcome λ_j is obtained with probability

$$p(\lambda_j) = \mathrm{Tr}\left(\Pi_j \rho\right). \tag{2.4}$$

The eigenvalues just represent an arbitrary labelling of the measurement outcomes, so a measurement can alternatively be represented by a set of orthogonal projection operators $\{\Pi_j\}$ that sum to the identity $\sum_j \Pi_j = I$, which is sometimes known as a *Projection Valued Measure (PVM)*.[12]

Two PVMs $A = \{\Pi_j\}$ and $B = \{\Pi'_j\}$ can be measured together if and only if each of the projectors commute, i.e. $\Pi_j \Pi'_k = \Pi'_k \Pi_j$ for all j and k. If this is the case then $\Pi_j \Pi'_k$ is also a projector and $\sum_{jk} \Pi_j \Pi_k = I$. Therefore, one way of performing the joint measurement is to measure the PVM $C = \{\Pi''_{jk}\}$ with projectors $\Pi''_{jk} = \Pi_j \Pi'_k$ and, upon obtaining the outcome (jk), report the outcome j for A and k for B. This fine graining procedure can be iterated by adding further commuting PVMs and forming the product of their elements with those of C. The procedure terminates when the resulting PVM is as fine grained as possible and this will happen when it consists of rank-1 projectors onto the elements of an orthonormal basis. The outcome of any other commuting PVM is determined by coarse graining the projectors onto the orthonormal basis elements.

Therefore, in quantum theory, we can take the sets $\Omega_{\mathcal{B}}$ that generate the betting contexts $\mathcal{B}$ to consist of the elements of orthonormal bases. An event $E \in \mathcal{B}$ is then a subset of the basis elements and corresponds to a projection operator $\Pi_E = \sum_{|\psi\rangle \in E} |\psi\rangle\langle\psi|$. The Boolean operations on $\mathcal{B}$ can be represented in terms of these projectors as

- Conjunction: $G = E \text{ AND } F \;\Rightarrow\; \Pi_G = \Pi_E \Pi_F$.
- Disjunction: $G = E \text{ OR } F \;\Rightarrow\; \Pi_G = \Pi_G + \Pi_F - \Pi_G \Pi_F$, which reduces to $\Pi_G = \Pi_G + \Pi_F$ when $E \cap F = \emptyset$.
- Negation: $G = \text{NOT } E \;\Rightarrow\; \Pi_G = I - \Pi_E$.

From the Dutch book argument applied within a betting context, we have that our degrees of belief should be represented by a set of probability measures $p(E|\mathcal{B})$ satisfying

- For any event $E \subseteq \Omega_{\mathcal{B}}$, $p(E|\mathcal{B}) \geq 0$.

[12] More generally, we could work with Positive Operator Valued Measures (POVMs) or sets of consistent histories, but this would not substantially change the arguments of this essay.

- For the certain events $\Omega_{\mathcal{B}}$, $p(\Omega_{\mathcal{B}}|\mathcal{B}) = 1$.
- For disjoint events within the same betting context $E, F \subseteq \Omega_{\mathcal{B}}$, $E \cap F = \emptyset$, $p(E \cup F|\mathcal{B}) = p(E|\mathcal{B}) + p(F|\mathcal{B})$.

The Born rule is an example of such an assignment, and in this language it takes the form

$$p(E|\mathcal{B}) = \text{Tr}\,(\Pi_E \rho)\,. \tag{2.5}$$

The Born rule also has the property that the probability only depends on the projector associated with an event, and not on the betting context that it occurs in. For example, in a three dimensional Hilbert space, consider the betting contexts $\Omega_{\mathcal{B}} = \{|0\rangle, |1\rangle, |2\rangle\}$ and $\Omega_{\mathcal{B}'} = \{|+\rangle, |-\rangle, |2\rangle\}$, where $|\pm\rangle = \frac{1}{\sqrt{2}}(|0\rangle \pm |1\rangle)$. The Born rule implies that $p(\{|2\rangle\}|\mathcal{B}) = p(\{|2\rangle\}|\mathcal{B}')$ and also that $p(\{|0\rangle, |1\rangle\}|\mathcal{B}) = p(\{|+\rangle, |-\rangle\}|\mathcal{B}')$ because, in each case, the events correspond to the same projectors. The Dutch book argument alone does not imply this because it does not impose any constraints across different betting contexts.

A probability assignment is called *noncontextual* if $p(E|\mathcal{B}) = p(F|\mathcal{B}')$ whenever $\Pi_E = \Pi_F$. Gleason's theorem [34] says that, in Hilbert spaces of dimension 3 or larger, noncontextual probability assignments are exactly those for which there exists a density operator ρ such that $p(E|\mathcal{B}) = \text{Tr}\,(\Pi_E \rho)$, i.e. they must take the form of the Born rule. Therefore, the Born rule follows from the conjunction of the Dutch book constraints and noncontextuality, at least in Hilbert spaces of dimension 3 or greater.

References

1. J.A. Wheeler, in *Proceedings of the 3rd International Symposium on Foundations of Quantum Mechanics in the Light of New Technology*, ed. by S. Kobayashi, H. Ezawa, Y. Murayama, S. Nomura (Physical Society of Japan, Tokyo, 1990), pp. 354–368
2. J.A. Wheeler, *in Problems in the Foundations of Physics: Proceedings of the International School of Physics "Enrico Fermi", Course LXXII*, ed. by G. Toraldo di Francia (North-Holland, Amsterdam, 1979), pp. 395–492
3. G. Birkhoff, J. von Neumann, Ann. Math. **37**, 823 (1936)
4. F.J. Murray, J. von Neumann, Ann. Math. **37**, 116 (1936)
5. M. Rédei, S.J. Summers, Stud. Hist. Philos. Mod. Phys. **38**, 390 (2007)
6. M.S. Leifer, R.W. Spekkens, Phys. Rev. A **88**, 052130 (2013). Eprint arXiv:1107.5849
7. D. Gillies, *Philosophical Theories of Probability* (Routledge, New York, 2000)
8. A. Eagle (ed.), *Philosophy of Probability: Contemporary Readings* (Routledge, 2011)
9. B. de Finetti, Fundamenta Mathematicae **17**, 298 (1931)
10. F.P. Ramsey, *The Foundations of Mathemaics and Other Logical Essays*, ed. by R.B. Braithwaite (Routledge and Kegan Paul, 1931), pp. 156–198. Reprinted in [8]
11. R. Jeffrey, *Subjective Probability: The Real Thing* (Cambridge University Press, New York, 2004)
12. C. Howson, P. Urbach, *Scientific Reasoning: The Bayesian Approach*, 3rd edn. (Open Court, 2005)
13. D. Lewis, *Studies in Inductive Logic and Probability*, ed. by R.C. Jeffrey (University of California Press, 1980). Reprinted with postscript in [8]

14. C.M. Caves, C.A. Fuchs, R. Schack, Phys. Rev. A **65**, 022305 (2002)
15. C.A. Fuchs, J. Mod. Opt. **50**, 987 (2003)
16. C.A. Fuchs, *Phy. Can.***66**, 77 (2010). Eprint arXiv:1003.5182
17. C.A. Fuchs, QBism, perimeter quantum Bayesianism (2010). Eprint arXiv:1003.5209
18. D. Deutsch, Proc. R. Soc. Lond. A **455**, 3129 (1999). Eprint arXiv:quant-ph/9906015
19. D. Wallace, Stud. Hist. Philos. Mod. Phys. **38**, 311 (2007). Eprint arXiv:quant-ph/0312157
20. D. Wallace, *Many Worlds? Everett, Quantum Theory, and Reality*, ed. by S. Saunders, J. Barrett, A. Kent, D. Wallace (Oxford University Press, 2010). Eprint arXiv:0906.2718
21. D. Wallace, *The Emergent Multiverse*, (Oxford University Press, 2012)
22. I. Pitowsky, Stud. Hist. Philos. Mod. Phys. **34**, 395 (2003). Eprint arXiv:quant-ph/0208121
23. B. de Finetti, *Philosophical Lectures on Probability*. Synthese Library vol. 340 (Springer, 2008)
24. E.T. Jaynes, *Probability Theory: The Logic of Science* (Cambridge University Press, 2003)
25. D. Kahneman, *Thinking, Fast and Slow* (Penguin, 2012)
26. C.A.B. Smith, J. R. Stat. Soc. B Met. **23**, 1 (1961)
27. L.J. Savage, *The Foundations of Statistics*, 2nd edn. (Dover, 1972)
28. H. Greaves, W. Myrvold, *Many Worlds?: Everett, Quantum Theory, and Reality*, ed. by S. Saunders, J. Barrett, A. Kent, D. Wallace (Oxford University Press, 2010), chap. 9, pp. 264–306
29. J. Barrett, Phys. Rev. A **75**, 032304 (2007). Eprint arXiv:quant-ph/0508211
30. S. Kochen, E.P. Specker, J. Math. Mech. **17**, 59 (1967)
31. W.H. Zurek, Phys. Rev. Lett. **90**, 120404 (2003). Eprint arXiv:quant-ph/0211037
32. W.H. Zurek, Phys. Rev. A **71**, 052105 (2005). Eprint arXiv:quant-ph/0405161
33. FQXi comment thread for this essay (2013). http://fqxi.org/community/forum/topic/1821
34. A.M. Gleason, J. Math. Mech. **6**, 885 (1957)
35. J.S. Bell, Rev. Mod. Phys. **38**, 447 (1966)
36. A. Chakravartty, *The Stanford Encyclopedia of Philosophy*, ed. by E.N. Zalta (2014), Spring 2014 edn. URL: http://plato.stanford.edu/archives/spr2014/entries/scientific-realism/
37. J. Ladyman, *Understanding Philosophy of Science* (Routledge, 2002)

Chapter 3
It from Qubit

Giacomo Mauro D'Ariano

Abstract In this essay I will embark on the venture of changing the realist reader's mind about the informational viewpoint for physics: "It from Bit". I will try to convince him of the amazing theoretical power of such paradigm. Contrarily to the common belief, the whole history of physics is indeed a winding road making the notion of "physical object"—the "It"—fade away. Such primary concept, on which the structure of contemporary theoretical physics is still grounded, is no longer logically tenable. The thesis I advocate here is that the "It" is emergent from pure information, an information of special kind: quantum. The paradigm then becomes: "It from Qubit". Quantum fields, particles, space-time and relativity simply emerge from countably infinitely many quantum systems in interaction. Don't think that, however, we can cheat by suitably programming a "simulation" of what we see. On the contrary: the quantum software is constrained by very strict rules of topological nature, which minimize the algorithmic complexity. The rules are: locality, unitarity, homogeneity, and isotropy of the processing, in addition to minimality of the quantum dimension. What is amazing is that from just such simple rules, and without using relativity, we obtain the Dirac field dynamics as emergent.

It is not easy to abandon the idea of a universe made of matter and embrace the vision of a reality made of pure information. The term "information" sounds vague, spiritualistic, against the attitude of concreteness that a scientist should conform to. We are all materialistic in the deep of our unconscious, we believe in "substance", and the idea of matter made of information (and not viceversa), seems inspired by a New-Age religion. It reminds us the immaterialism of bishop Berkeley. Software without

---

The following dissertation is a minimally updated version of the original essay presented at the FQXi Essay Contest 2013 "It From Bit or Bit From It". A short summary of the follow-ups and main research results is given in the Postscriptum.

G.M. D'Ariano (✉)
Dipartimento di Fisica dell'Università degli Studi di Pavia, via Bassi 6, 27100 Pavia, Italy
e-mail: dariano@unipv.it

G.M. D'Ariano
Istituto nazionale di Fisica Nucleare, Gruppo IV, Sezione di Pavia, Italy

A. Aguirre et al. (eds.), *It From Bit or Bit From It?*,
The Frontiers Collection, DOI 10.1007/978-3-319-12946-4_3

hardware? Nonsense. Information about what? Whose information? A subjective information? We cannot give-up objectivity of science!

I will try to convince you that we can reconcile objectivity with subjectivity by embracing a more pragmatic kind of realism, based on what we observe and not on what we believe is out there. In the scientific process we are easily lead to consider as "ontic" entities that are instead only theoretical notions. We must separate what should be taken as "objective" from what is element of the theory, and define precisely the boundary between theory and observation. Science must make precise predictions about what everybody agree on: the observed facts, the "events".

"Informationalism": A Realistic Immaterialism

Quantum Mechanics has taught us that we must change our way of thinking about "realism", and that this cannot be synonymous of "materialism". Likewise objectivity should not be confused with the availability of a physical picture in terms of a "visible" mechanism. We must specify which notions have the objectivity status, and describe the experiment in terms of them. What matters is our ability of making correct predictions, not of "describing what is out there as it is"—a nonsense, since nobody can check it for us. We only need to describe logically and efficiently what we see, and for such purpose we conveniently create appropriate "ontologies", which nonetheless are useful tools for depicting mechanisms in our mind.

Why we should bother changing our way of looking at reality? Because the old matter-realistic way of thinking in terms of particles moving around and interacting on the stage of space-time is literally blocking the progress of theoretical physics. We know that we cannot reconcile general relativity and quantum field theory, our two best theoretical frameworks. They work astonishingly well within the physical domain for which they have been designed. But the clash between the two is logically solved only if we admit that they are not both correct: at least one of them must hold only approximately, and emerge from an underlying more fundamental theory. Which one of the two? The answer from "It from Qubit" is: relativity theory! Indeed, the informational paradigm shows its full power in solving the clash between the two theories (at least if we restrict to special relativity), with relativity derived as emergent from quantum theory of interacting systems—*qubits* at the very tiny Planck scale.

A description of a reality emerging from pure software would not provide a good theory if we were allowed to adjust the "program" to make it work. The "subroutines" must stringently derive from few very general principles, corresponding to minimizing the algorithmic complexity: this is the new "elementarity" notion that will substitute the corresponding one in particle physics. What is now astonishing is that few simple topological principles—locality, homogeneity, and isotropy, unitarity, linearity, and minimality of quantum dimension—lead to the Dirac field theory, without assuming relativity. The only great miracle here, as it always happens with physics, is the amazing power of mathematics in describing the world. But is it really a miracle?

The Notion of Physical Object Is Untenable

Matter is not made of matter

Hans Peter Dürr

In physics we are accustomed to think in terms of physical "objects" having "properties" (location, speed, color,...), the value of each property depending on the object's "state". The object considered as a "whole", is taken as the sum of is "parts". The dynamics accounts for the evolution of the state, or equivalently of the properties of the object, and is described in terms of "free" dynamics for each part, along with "interactions" between the parts, each part retaining its individuality, namely being itself an object with its own properties. This bottom-up approach is called "reductionism", and is opposed to "holism", according to which the properties of the whole cannot be understood in terms of the properties of the parts. Holism is commonly contrasted to the mechanical "clockwork" picture of nature inherited from the scientific revolution, emphasizing it as a motivation for integrating top-down approaches. One of the unexpected features of quantum mechanics is that it incorporates a form of holism absent from classical physics. In addition, the theory entails "complementarity", namely the existence of incompatible properties that cannot be shared by an object in any possible state, nevertheless providing different kinds of information about it. The state of the object generally does not correspond to a precise value of the property, but provides the probability distribution of values of each property.

Reconciling Holism with Reductionism. Quantum theory entails a strong instance of holism, with the existence of properties of the whole that are incompatible with any property of the parts. Correspondingly, there are states of the whole with determinate values of a property of the whole, but having no determinate value of any property of the parts. Thus, differently from classical mechanics, we have the seemingly paradoxical situation that we can have perfect knowledge of the whole having no knowledge of the parts. Such holistic states of the whole describe correlations between properties of the parts that cannot be interpreted as shared randomness, namely they do not correspond to a joint probability distribution of random values of the properties of the parts. This is what we call "quantum non-locality", and it is signaled by the violation of the celebrated Bell's bound for shared randomness [1], which has been breached in numerous experiments in quantum optics and particle physics.

The holism of quantum theory has resulted in the popular credo that quantum theory is logically inconsistent with the bottom-up approach of physics. On the contrary, the structure of the theory is fully consistent with it. How the theory reconciles with the bottom-up approach? The answer relays on the fact that the theory satisfies the principle of "local discriminability" [2, 3], namely the possibility of discriminating between any two states of the whole by performing only observations on the parts. This means that we can still observe a holistic reality in a reductionistic way by observing only the parts of the whole.

The Bell Test Supports a Deeper Epistemological Realism. The Bell result changes dramatically our way of looking at reality, and for this reason it shows the epistemological power of physics in guiding our knowledge well beyond the mere appearance. We can tell whether the deep conceptual framework of the theory is in focus, and be well aware of its reliability and theorizing perspectives, a step essential to objectivity. At first glance, Bell's theorem seems to be against realism, for the inescap able holism that proves the inextricable interconnectedness of parts that blurs their individual images. Instead, the Bell test supports a deeper epistemological realism, providing a strong positive case for our ability to go beyond the appearance. Things are not the way we naively believed they are: realism cannot mean that we should be able to see sharply defined parts the way we believe they exist out there. Contrarily to what Einstein thought, such an intrinsic unsharpness is not the incapability of quantum theory to go beyond the veil that blurs our observation: it is the way things are. The lesson spelled loud and clear by the Bell theorem is that we should trust observations, even against our intuition, and ground our knowledge on the logic of the experiment, focusing theoretical predictions on what we actually observe. In a word: being operationalist.

The Plato's Cave and the Shadows of Physical Ontologies. We are like the prisoners in Plato's cave, who can see objects only through the shadows they cast. The "true" object may have properties in addition to what we see, e.g. three dimensional shape and color—properties that are seemingly irrelevant for the casted shadows. The detractor of operationalism would say that the doctrine rejects as unphysical those hidden variables with no immediate empirical consequences. However, pragmatically such restriction should be taken only as long as the hidden variables have no additional explanatory power, e.g. in describing the dynamics of the shadows overlapping each other on the walls of the cave. We can create a three-dimensional ontology corresponding to the shadows, but we should not forget that this is an explanatory tool, not "what is really out there". The ontology can be extremely powerful in describing a large number of different phenomena, as it is the case of the modern notion of atom, on which the whole chemistry relies, and which allows us understanding a great deal of physics. Nowadays we can almost "see" the atoms using a tunnel- effect microscope, even though we shouldn't forget that these images are just a suitable mathematical representation of electric signals. Ernst Mach was stubbornly against the idea of atoms, but he was proven wrong.

The Elementary-Particle Ontology. An evolution of the notion of atom is the modern concept of elementary particle, which has marked the greatest successes of modern physics. Unfortunately, we have not onlysuccesses, but also failures in explaining relevant phenomena—e.g. gravity or dark matter and other astrophysical observations—phenomena that even a reasonable revision of the particle notion seems unable to explain.

An ontology that works perfectly well in accounting for a large class of phenomena may later be proved having not the same power in explaining other phenomena, e.g. those occurring at scales that are much larger or much smaller than those where the ontology is successful. Ultimately the ontology may turn out to be even logically inconsistent with the theoretical framework itself: later, a new more powerful ontology will emerge, which can account for mechanisms within a much larger physical domain, and without suffering the logical inconsistencies of the old ontology.

We must always keep in mind that the motivations for adopting the new ontology must always be its additional explanatory power in accounting for the behavior of the observed shadows on the cave walls, and, more important, the logical solidity and consistency of the theoretical principles embodied by the ontology. Unfortunately, some colleagues followers of Einstein's realism got so fond of the Plato's cave paradigmatic tale, to the extent that they believe that quantum mechanics only describes the shadows on the cave walls, whereas they are convinced that there exists a veiled reality made of particles like three-dimensional marbles: this is what they call the "true reality". But here the Bell's theorem comes to help us, proving that, whatever outside the cave the object are made of, they cannot be constituted of "parts" of which we can have perfect knowledge in all cases. Quantum nonlocality is not a feature of the shadows only: it holds for any possible object projecting the shadow. This is the amazing epistemological power of physics.

The Evaporation of the Notion of Object

Quine in his *Whither Physical Objects?* [4] made a thorough attempt to arrive at a very comprehensive concept of "object", but he end up with a progressive evaporation of the notion, from the "body", toward "space-time region", up to mere "set of numerical coordinates" with which he ends.

What is a "physical object"? Independently on the specific context, an object must be located in space and time. Its persistence through time is a fundamental feature to grant its individuality. What if we have two identical objects A and B that disappear and suddenly reappear somewhere else? How can we know which one is A and which is B? This is exactly what happens with identical quantum particles, which are literally indistinguishable. And, indeed, they cannot be followed along their trajectories, even in principle. "Particles", i.e. "small parts", are the minimum "parts" of which every material object is made up. But can we consider particles as objects themselves?

Take the "atom" as the ancestor notion of particle. Since its birth with Democritus and Leucippus, the idea of atom was devised to solve precisely the problem of individuality of objects. Is an object something different from the stuff it is made of? Heraclitus said that "we could not step twice into the same river", to emphasize that the river is never the same water, contrarily to appearance. The river is not the collection of water drops: it is a bunch of topological invariants in the landscape: the two sides, the flow of water in between. Thus the notion of physical object resorts to a set of invariants. And the atoms are invariants, eternal entities within the river flow.

The Theseus' Ship Paradox and Teleportation: "It" Becomes "State". In a popular tale Plutarch raised the following paradox: the Theseus' ship was restored completely, by replacing all its wooden parts. After the restoration, was it the same ship? The problem of the theseus' ship can be posed more dramatically in modern terms, using the thought experiment in which a human is teleported between two places very far apart, e.g. Earth and a planet of Alpha Centauri. From quantum theory we know the basic principles of teleportation. Each atom, electron, proton, neutron, etc. of the human body undergoes a quantum measurement that completely destroys its quantum state. A huge file containing all measurement outcomes is sent to the arrival place (to cover the distance between the two planets it will take 4.37 years traveling at the speed of light). At the arrival the quantum state is rebuild over local raw matter.

Technically a so-called entangled resource is needed, namely a bunch of previously prepared particle states of the same kind of those used to experimentally prove violation of the Bell's bound. According to quantum theory the protons (neutrons, electrons, etc.) at the departure point are indistinguishable, even in principle, from those at the arrival point: matter is the same everywhere. The quantum measurement while destroying the quantum state of the human's molecules, literally kills the person, reducing him to raw matter. Then, the rebuilding of the human at the arrival is made by re-preparing the matter available there in the same original state that the human had at the departure point: teleportation literally resurrects the human. The question now is: are the human before and the human after teleportation the same individual? The two are indeed perfectly indistinguishable: they are made of the same matter, and even share the same thoughts, since the molecules of the brain are in the same physical state as they were before teleportation (indeed, the teleported guy will feel to be the same individual, and had experienced just a sudden change of his surrounding).

What is then the teleported human? He is certainly not identifiable with his constituent matter: matter is everywhere the same. The human is the shape along with all the properties of the matter that is made of. Apart from a space translation, the human is a "state" of matter—a very complicate state indeed, involving many particles. But with this reasoning we have reached an inconsistency with the original notion of object, since the state is not the object itself, but it is a catalog of all its properties. This means that what we considered an object was instead a "state"—as the shape of the river, the shape of the Theseus' ship—whereas the physical objects are now the particles, the stuff.

Quantum Field Theory: The Particle Becomes a State. We enter now quantum field theory, and what we discover? We realize that, differently from the non relativistic quantum mechanics, particles are themselves states of something else: the quantum field. Thus, electrons are states of the electron field, photons are states of the electromagnetic fields, neutrinos of the neutrino field, and so on. The process of demoting particles to states and introducing the notion of quantum field as the new "object" for such states is known as "second quantization".

The Field is Not an "Object". But is now the field an object in the usual sense? Not at all. The field is everywhere. And it is not made of matter: its states are. What is it then? It is a collection of infinitely many quantum systems. But the "quantum system" is an abstract notion: it is an immaterial support for quantum states, exactly in the same fashion as the "bit" in computer science is the abstract system having the two states 0 and 1. The analogous system of the bit in quantum theory is the "qubit", having not only the two states 0 and 1, but also all their superpositions, corresponding to the possibility of having complementary properties which are absent in classical computer science. Therefore, we are left with states of qubits, namely pure quantum software: objects, matter, hardware, completely became vaporized.

It from Qubit: Space-Time Emerging from a Web of Interactions

A Game on the Web. Consider the following game on the web. There is an unbounded number of players: Alice, Bob, Carol, David, Eddie, ... Each player has the same identical finite set $S = \{e, h_1, h_2, \ldots, h_M, h_1^{-1}, h_2^{-1}, \ldots, h_M^{-1}\}$ of colored buttons to press. When pressing button e one connects with himself, and experiences audio feedback. When pressing button h_1 Alice speaks with Bob, whereas when Bob presses the button h_1^{-1} he speaks to Alice. If Alice presses h_1 and Bob presses h^{-1} both will experience audio feedback. After trying many connections, Alice realizes that when she presses h_1^{-1} and Bob presses h_2 connecting to Carol, and Carol presses h_3, all of them experience audio feedback, meaning that Carol is connected back to Alice. The same happens if anybody else presses h_1, and the connected player presses h_2, and the third connected player presses h_3: the same feedback loop holds starting from any player, namely from the network perspective all players are perfectly equivalent. Also the feedback delay in the two- person round-trip communication is the same for every player and for every pressed button: it is $2t_P$. Then the delay for each feedback loop is a multiple of t_P, e.g. the delay of the Alice-Bob-Carol-Alice loop is $3t_P$. Each players doesn't know where the other players are: they can only try to figure it out from the feedback loop structure and the delays. It is easy to realize that the above structure is that of a group, which we will call G : e is the identity element, h_j the group generators, h_j^{-1} the respective inverses, whereas the feedback loops are relations among group elements, e.g. $h_3h_2h_1 = e$, or $h_2h_1 = h_3^{-1}$. Each player corresponds to an element of the group. The fact that all players are equivalent corresponds to the homogeneity of the group network (this network is precisely a Cayley graph of the group). Thus, by playing the game and by knowing that the network is homogeneous, we come out with a group G which is given by the so-called group presentation, i.e. via generators and relators. Generally even though the group is finitely generated, it grows unbounded. This is the case, for example, of a lattice, as those of crystals. For example, in the simple-cubic lattice there are only three generators (the translations along x, y, z), and along with their respective inverses they make a total of six elements, corresponding to the coordination number of the lattice. The time-delay of the feedback loops is a way of measuring the distance between the players: it is a metric for the group: the so-called "word-metric" (the numbers of letters of the word denoting the group multiplication, e.g. for $h_3h_2h_1$ the length is 3). From the feedback loops we figure out the shape of the network, e.g. a simple-cubic lattice. We then imagine the network immersed in the usual Euclidean space R^3. There is, however, a mismatch between the distances measured in R^3 and those measured with the word-metric: they are exactly proportional when measured along a fixed direction, but the proportionality constant differs depending on direction, e.g. it is 1, $\sqrt{2}$, or $\sqrt{3}$ if measured along the sides, the face-diagonals or the main diagonals of the cubes, respectively. This mismatch has been noted by Weyl [5], who argued that we cannot have a continuous geometry emerging from a discrete one, since we could never get the irrational numbers as $\sqrt{2}$ or $\sqrt{3}$ coming from the Pythagoras' theorem. Then, we cannot immerse the lattice in R^3 by preserving the metric, since the word-metric and the Euclidean metric cannot be matched. In mathematical terms we say that the lattice cannot be isometrically embedded in R^3. But here a new outstanding branch of mathematics comes to help: the geometric-group theory of Gromov [6]. It states that we only need a quasi-isometric embedding, namely the two metrics should match modulo additive and multiplicative constants. (Geometric-group connects algebraic properties of groups with topological and geometric properties of spaces on which these groups act).

Now you would ask: why such a construction for having space-time as emergent? The answer is that we want to have space-time and relativity emerging from just quantum systems in interactions. In the game on the web, the players $g \in G$ label the quantum systems $\psi(g)$, which is a vector/spinor quantum field evaluated at $g \in G$. The player connections $h_i \in S$ label their local interactions in terms of transitions matrices A_h. The whole quantum network of systems is a Quantum Cellular Automaton, our quantum software. The single-step of the run is described by the unitary operator [7]

$$A = \sum_{h \in S} T_h \otimes A_h$$

where T_h is a unitary representation of the group G. Thanks to its quantum nature, the automaton physically achieves the quasi-isometric embedding, and on the large scale we recover the relativistic quantum field theory.

The Quantum Cellular Automata. One can ask: what is the minimal field vector dimension $\mathcal{S}$ of a nontrivial automaton quasi-isometrically embeddable in $R^3$and isotropic? For $\mathcal{S} = 1$ the automaton is trivial. For $\mathcal{S} = 2$ it turns out that there are two automata that are reciprocally connected by chirality (all results that follow have been presented in the joint work with Perinotti [7]). The groups that are quasi-isometrically embeddable in R^3 must be commutative, and these are the Bravais lattices, and the only lattice that achieve unitarity and isotropy is the BCC (body cubic centered). The eigenvalues of A have unit modulus, and their phases as a function of the wave-vector $\mathbf{k}$ in the Brillouin zone are the dispersion relations. For $|\mathbf{k}| \ll 1$ (the so-called relativistic regime) the two automata approaches the Weyl equation. Coupling such Weyl automata in the only possible localized way, one gets two different automata with $\mathcal{S} = 4$ that are reciprocally connected by the CPT symmetry. Thus, the CPT symmetry is broken, and is recovered in the relativistic limit, where both automata become the Dirac equation, with the rest-mass being the coupling constant. Therefore, the simplest cellular automata satisfying unitarity, locality, homogeneity, and isotropy are just those achieving the Weyl and Dirac equations in the limit of small wave-vectors. For general $\mathbf{k}$ the automata can be regarded as a theory unifying scales from Planck to Fermi, with Lorentz covariance distorted [13] *a la* Amelino-Camelia [9, 10] and Smolin/Magueijo [11, 12], i.e. with additional invariants in terms of energy and length scales. They exhibit relative locality [14], namely event coincidence depending on the observer and on the momentum of the observed particles. The generalized energy-momentum Lorentz trans- formations are those that leave the dispersion relations invariant [13]. Thus, relativistic quantum field theory is obtained without assuming relativity, as a theory emergent at large scales from a more fundamental theory of information processing. This has also been shown in Ref. [13] for the one-dimensional Dirac automaton earlier derived by heuristic arguments [15]. For technical details of the Dirac automata in R^d with $d = 1, 2, 3$ the reader can see Refs. [7, 13, 16].

The Many Bonuses of the It-from-Qubit

In addition to emergence of relativistic quantum field and space-time without assuming relativity, the quantum automaton theory has a number of very desirable features that are not possessed by quantum field theory. The theory is quantum ab-initio, and is the natural scenario for the holographic principle, two dreamy features for a microscopic theory of gravity a la Jacobson [17] and Verlinde [18]. It extends field theory by including localized states and measurements, solving the issue of localization of quantum field theory. It has no violation of causality and no superluminal tail of the wave-function. It is computable and is not afflicted by any kind of divergence. Its dynamic is stable, allowing analytical evaluations of the evolution for long times, a feature that is crucial for deriving observable phenomenology. Despite its simplicity it leads to unexpected interesting predictions, e.g. it anticipates a bound for the rest-mass for the Dirac particle, where the particle behaves as a mini black-hole, without using general relativity, only as a consequence of unitarity [16].

The predicted violation of Lorentz covariance and space-isotropy affect physics at huge energies, many order of magnitude above that of ultra-high-energy cosmic rays. Planck-scale effects are possibly visible from light coming from quasars at the boundary of the universe [19, 20].

The quantum nature of the automaton is crucial for the emergence of space-time, since continuous isotropy and all continuous symmetries are recovered from the discrete ones in the relativistic limit thanks to quantum interference between paths [21] (Lorentz covariance from classical causal networks conflicts with homogeneity, and needs a random topology [22]). The classical dynamics also emerges from the automaton, with the particle trajectories being the "typical paths" of narrow-band superpositions of single-excitations, whereas the field Hamiltonian is derived from the unitary operator A [16].

Postscriptum

All predictions contained in this Essay has been later derived, and are now available in technical papers. The reader should look at Ref. [7] and the new Refs. [23, 25, 26]. The main result is contained in manuscript [7], entitled "Derivation of the Dirac equation from informational principles". There it is proved the remarkable result that from the only general assumptions of locality, homogeneity, isotropy, linearity and unitarity of the interaction network, only two quantum cellular automata follow that have minimum dimension two, corresponding to a Fermi field. The two automata are connected by CPT, manifesting the breaking of Lorentz covariance. Both automata converge to the Weyl equation in the relativistic limit of small wave-vectors, where Lorentz covariance is restored. Instead, in the ultra- relativistic limit of large wave-vectors (i.e. at the Planck scale), in addition to the speed of light one has extra invariants in terms of energy, momentum, and length scales. The resulting distorted Lorentz covariance belongs to the class of the *Doubly Special Relativity* of Amelino-Camelia/Smolin/Magueijo. Such theory predicts the phenomenon of *relative locality*, namely that also coincidence in space, not only in time, depends on the reference frame. In terms of energy and momentum covariance is given by the group of transformations that leave the automaton dispersion relations unchanged. Via Fourier transform one recovers a space-time of quantum nature, with points in superposition. All the above results about distorted Lorentz covariance are derived in Ref. [13].

The Weyl QCA is the elementary building block for both the Dirac and the Maxwell field. The latter is recovered in the form of the de Broglie neutrino theory of the photon. The Fermionic fundamental nature of light follows from the minimality of the field dimension, which leads to theory Boson as an emergent notion [26].

The discrete framework of the theory allows to avoid all problems that plague quantum field theory arising from the continuum, including the outstanding problem of localization. Most relevant, the theory is quantum ab initio, with no need of quantization rules.

References

1. J.S. Bell, *Speakable and Unspeakable in Quantum Mechanics* (Cambridge University Press, Cambridge, 1987)
2. G.M. D'Ariano, in *Philosophy of Quantum Information and Entanglement*, ed. by A. Bokulich, G. Jaeger (Cambridge University Press, Cambridge, 2010), p. 85
3. G. Chiribella, G.M. D'Ariano, P. Perinotti, Phys. Rev. A **84**, 012311 (2011)
4. W.O. van Quine, Wither Physical Objects, in *Essays in Memory of Imre Lakatos*, ed. by R.S. Cohen, P.K. Feyerabend, M.W. Wartofsky (Reidel, Dordrecht, 1976)
5. H. Weyl, *Philosophy of Mathematics and Natural Sciences* (Princeton University Press, Princeton, 1949)
6. P. de La Harpe, *Topics in Geometric Group Theory* (University of Chicago Press, Chicago, 2000)

7. G.M. D'Ariano, P. Perinotti, (2013) arXiv: 1306.1934
8. M. Kapovich, Cayley graphs of finitely generated infinite groups quasi-isometrically embeddable in R^3, http://mathoverflow.net/questions/130994 version: 2013-05-17
9. G. Amelino-Camelia, Int. J. Mod. Phys. D **11**, 35 (2002)
10. G. Amelino-Camelia, T. Piran, Phys. Rev. D **64**, 036005 (2001)
11. J. Magueijo, L. Smolin, Phys. Rev. D **88**, 190403 (2002)
12. J. Magueijo, L. Smolin, Phys. Rev. D **67**, 044017 (2003)
13. A. Bibeau-Delisle, A. Bisio, G. M. D'Ariano, P. Perinotti, A. Tosini, (2013) arXiv:1310.6760
14. G. Amelino-Camelia, L. Freidel, J. Kowalski-Glikman, L. Smolin, (2011) arXiv:1106.0313
15. G.M. D'Ariano, AIP Conf. Proc. **1327**, 7 (2011)
16. G.M. D'Ariano, Phys. Lett. A **376**, 697 (2011)
17. T. Jacobson, Phys. Rev. Lett. **75**, 1260 (1995)
18. E. Verlinde, J. High Energy Phys. **29**, 1 (2011)
19. W.A. Christiansen, D.J.E. Floyd, Y.J. Ng, E.S. Perlman, (2009) arXiv:0912.0535
20. F. Tamburini, C. Cuofano, M. Della Valle, R. Gilmozzi, (2011) arXiv:1108.6005
21. G.M. D'Ariano, in FQXi Essay Contest: *Questioning the Foundations: Which of Our Basic Physical Assumptions Are Wrong?* (2012)
22. L. Bombelli, J.H. Lee, D. Meyer, R. Sorkin, Phys. Rev. Lett. **59**, 521 (1987)
23. A. Bisio, G. M. D'Ariano, A. Tosini, (2012) arXiv:1212.2839
24. G.M. D'Ariano, F. Manessi, P. Perinotti, A. Tosini, Int. J. Mod. Phys. A **29**, 1430025 (2014)
25. A. Bisio, G.M. D'Ariano, A. Tosini, Phys. Rev. A **88**, 032301 (2013)
26. A. Bisio, G. M. D'Ariano, P. Perinotti arXiv:1407.6928

Chapter 4
Drawing Quantum Contextuality with 'Dessins d'enfants'

Michel Planat

Abstract In the standard formulation of quantum mechanics, there exists an inherent feedback of the measurement setting on the elementary object under scrutiny. Thus one cannot assume that an 'element of reality' prexists to the measurement and, it is even more intriguing that unperformed/counterfactual observables enter the game. This is called quantum contextuality. Simple finite projective geometries are a good way to picture the commutation relations of quantum observables entering the context, at least for systems with two or three parties. In the essay, it is further discovered a mathematical mechanism for 'drawing' the contexts. The so-called 'dessins d'enfants' of the celebrated mathematician Alexandre Grothendieck feature group, graph, topological, geometric and algebraic properties of the quantum contexts that would otherwise have been 'hidden' in the apparent randomness of measurement outcomes.

Introduction

The motivation for being interested in the topic of quantum contextuality dates back the celebrated Bohr-Einstein dialogue about the fundamental nature of quantum reality. The first sentence of the Einstein Podolski Rosen (EPR) paper [1] is as follows

> If, without in any way disturbing a system, we can predict with certainty (i.e., with probability equal to unity) the value of a physical quantity, then there exists an element of physical reality corresponding to that physical quantity.

and the last sentence in Bohr's [2] reply is

> I should like to point out, however, that the named criterion contains an essential ambiguity when it is applied to problems of quantum mechanics. It is true that in the measurements under consideration any direct mechanical interaction of the system and the measuring agencies is excluded, but a closer examination reveals that the procedure of measurements has an essential influence on the conditions on which the very definition of the physical

M. Planat (✉)
Institut FEMTO-ST, CNRS, 15 B Avenue des Montboucons, 25044 Besançon, France
e-mail: michel.planat@femto-st.fr

A. Aguirre et al. (eds.), *It From Bit or Bit From It?*,
The Frontiers Collection, DOI 10.1007/978-3-319-12946-4_4

quantities in question rests. Since these conditions must be considered as an inherent element of any phenomenon to which the term "physical reality" can be unambiguously applied, the conclusion of the above mentioned authors would not appear to be justified.

In a recent essay, taking into account the work of Bell about non-locality [3] and further important papers by Gleason, Kochen-Specker [4, 5] and Mermin [6, 7], I arrived at the conclusion that further progress about the elusive *elements of reality*, or rather the *elements of knowledge*, can be performed by resorting to Grothendieck's *dessins d'enfants* as summarized in the note [8] and the paper [9] intended to illustrate Wheeler's *from bit* perspective [10, 11]. In Grothendieck's words [12, Vol.1], [13]

The demands of university teaching, addressed to students (including those said to be advanced) with a modest (and frequently less than modest) mathematical baggage, led me to a Draconian renewal of the themes of reflection I proposed to my students, and gradually to myself as well. It seemed important to me to start from an intuitive baggage common to everyone, independent of any technical language used to express it, and anterior to any such language it turned out that the geometric and topological intuition of shapes, particularly two-dimensional shapes, formed such a common ground. This consists of themes which can be grouped under the general name of topology of surfaces or geometry of surfaces, it being understood in this last expression that the main emphasis is on the topological properties of the surfaces, or the combinatorial aspects which form the most down-to-earth technical expression of them, and not on the differential, conformal, Riemannian, holomorphic aspects, and (from there) on to complex algebraic curves. Once this last step is taken, however, algebraic geometry (my former love!) suddenly bursts forth once again, and this via the objects which we can consider as the basic building blocks for all other algebraic varieties. Whereas in my research before 1970, my attention was systematically directed towards objects of maximal generality, in order to uncover a general language adequate for the world of algebraic geometry, and I never restricted myself to algebraic curves except when strictly necessary (notably in etale cohomology), preferring to develop pass-key techniques and statements valid in all dimensions and in every place (I mean, over all base schemes, or even base ringed topoi...), here I was brought back, via objects so simple that a child learns them while playing, to the beginnings and origins of algebraic geometry, familiar to Riemann and his followers!

Dessins d'enfants (also known as bicolored maps) are bipartite graphs drawn on a smooth surface but they also possess manifold aspects. They are at the same time group theoretical, topological and algebraic objects and, as revealed by the author, they allow to stabilize the finite geometries attached to quantum contexts. There seems to exist a remarkable confluence between the so-called 'magic' configurations of quantum observables found to illustrate the no-go theorems à la Kochen-Specker and the symmetries obeyed by the algebraic extensions over the field of rational numbers, a subject briefly advocated by Grothendieck as

In the form in which Belyi states it, his result essentially says that every algebraic curve defined over a number field can be obtained as a covering of the projective line ramified only over the points 0, 1 and ∞. The result seems to have remained more or less unobserved. Yet it appears to me to have considerable importance. To me, its essential message is that there is a profound identity between the combinatorics of finite maps on the one hand, and the geometry of algebraic curves defined over number fields on the other. This deep result, together with the algebraic interpretation of maps, opens the door into a new, unexplored world - within reach of all, who pass by without seeing it.

In section "The Manifolds Traits of a 'Dessin D'enfant'", a brief account of the 'technology' of dessins d'enfants is provided. Section "Bell's Theorem with 'Dessins D'enfants'" addresses the relation between Bell's theorem and some dessins 'living' in the extension field $\mathbb{Q}(\sqrt{2})$. Sections "Kochen-Specker Theorem with 'Dessins D'enfants': Two Qubits" and "Kochen-Specker Theorem with 'Dessins D'enfants': Three Qubits" deal about dessins attached to the two- and three-qubit Kochen-Specker theorem about contextuality. Then, in section "Dessins D'enfants and Generalized Polygons", it is shown that generalized polygons $GQ(2, 2)$ and $GH(2, 2)$, and their corresponding driving dessins, encode the commutation relations of two- and three-qubit systems, respectively. As our last example, in section "A Dessin D'enfant for Six-Qudit Contextuality", a dessin related to the contextuality of the six-qudit system, described in [14], is displayed.

The Manifolds Traits of a 'Dessin D'enfant'

I will explain that quantum contexts can be drawn as Grothendieck's 'dessins d'enfants'. A 'true' dessin d'enfant is shown in Fig. 4.1. This section accounts for the mathematics of 'false' 'dessins d'enfants'. Of course, there are constraints that a child does not take care about: the 'dessins' in question are connected, they are bipartite with black and white points and they are also chosen to be 'clean', meaning that the valency of white vertices is ≤ 2. The last constraint can easily be removed—the valency of white vertices can be made arbitrary to correspond to an hypermap—although this is not necessary for our quantum topic. Doing this, a 'dessin' acquires a *topological genus* g (which quantifies the number of holes on the smooth surface where it is drawn) such that $2 - 2g = B + W + F - n$, where B, W, F and n stands for the number of black vertices, the number of white vertices, the number of faces and the number of edges, respectively.

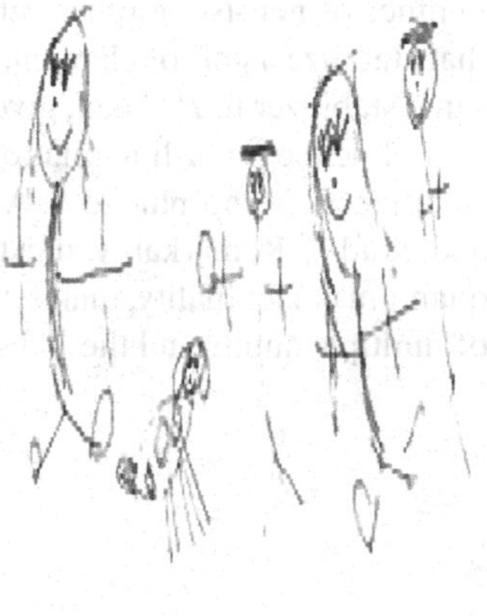

Fig. 4.1 A dessin d'enfant drawn at a kindergarten by a five-year old girl, http://www.parents.com/fun/arts-crafts/kid/decode-child-drawings/

Given a dessin $\mathcal{D}$ with n edges labeled from 1 to n, one can recover the *combinatorial information* by associating with it a two-generator permutation group $P = \langle \alpha, \beta \rangle$ on the set of labels such that a cycle of α (resp. β) contains the labels of the edges incident to a black vertex (resp. white vertex) and by computing a passport [15] in the form $[C_\alpha, C_\beta, C_\gamma]$, where the entry $C_i, i \in \{\alpha, \beta, \gamma\}$ has factors $l_i^{n_i}$, with l_i denoting the length of the cycle and n_i the number of cycles of length l_i.

Another observation made by Grothendieck is of utmost importance. The dessins are in one-to-one correspondence with conjugacy classes of subgroups of finite index of the triangle group $C_2^+ = \langle \rho_0, \rho_1, \rho_2 | \rho_1^2 = \rho_0 \rho_1 \rho_2 = 1 \rangle$. The existence of dessins with prescribed properties can thus be straightforwardly checked from a systematic enumeration of conjugacy classes of C_2^+. Note that enumeration becomes tedius for large dessins since the number of dessins grows exponentially with the number of their edges. To proceed with the effective calculations of a dessin, one counts the cosets of a subgroup of C_2^+ and determine the corresponding permutation representation P by means of the Todd-Coxeter algorithm implemented in an algebra software such as Magma.

Then, according to Belyi's theorem, a dessin may be seen as an *algebraic curve* over the rationals. Technically, the *Belyi function* corresponding to a dessin $\mathcal{D}$ is a rational function $f(x)$ of the complex variable x, of degree n, such that (i) the black vertices are the roots of the equation $f(x) = 0$ with the multiplicity of each root being equal to the degree of the corresponding (black) vertex, (ii) the white vertices are the roots of the equation $f(x) = 1$ with the multiplicity of each root being equal to the degree of the corresponding (white) vertex, (iii) the bicolored graph is the preimage of the segment [0, 1], that is $\mathcal{D} = f^{-1}([0, 1])$, (iv) there exists a single pole of $f(x)$, i.e. a root of the equation $f(x) = \infty$, at each face, the multiplicity of the pole being equal to the degree of the face, and, finally, (v) besides 0, 1 and ∞, there are no other critical values of f [9, 15]. This construction works well for small dessins $\mathcal{D}$ but it becomes intractable for those with a high index n; however a complex algebraic curve is associated to every $\mathcal{D}$.

Last but not least, in many cases, one may establish a bijection between notable point/line incidence geometries $\mathcal{G}_{\mathcal{D}}^i$ to a dessin $\mathcal{D}$, $i = 1, \ldots, m$ with m being the number of non-isomorphic subgroups S of the permutation group P of the dessin that stabilize a pair of elements. We ask that every pair of points on a line shares the same stabilizer in P. Then, given a subgroup S of P which stabilizes a pair of points, we define the point-line relation on $\mathcal{G}_{\mathcal{D}}$ such that two points will be adjacent if their stabilizer is isomorphic to S. A catalog of small finite geometries is given as Tables 1 and 2 of [9]. Remarkably, most geometries derived so far have been found to rely on quantum contextuality, that is, the points of a $\mathcal{G}_{\mathcal{D}}$ correspond to quantum observables of multiple qubits and the lines are mutually commuting subsets of them.

Bell's Theorem with 'Dessins D'enfants'

John Bell:

> First, and those of us who are inspired by Einstein would like this best, quantum mechanics may be wrong in sufficiently critical situations. Perhaps nature is not so queer as quantum mechanics. But the experimental situation is not very encouraging from this point of view... Secondly, it may be that it is not permissible to regard the experimental settings a and b in the analyzers as independent variables, as we did. We supposed them in particular to be independent of the supplementary variables λ, in that a and b could be changed without changing the probability distribution $\rho(\lambda)$ ... Apparently separate parts of the world would be deeply and conspirationaly entangled, and our apparent free will would be entangled with them. Thirdly, it may be that we have to admit causal influences do go faster than light... Fourthly and finally, it may be that Bohr's intuition was right—in that there is no reality below some 'classical' 'macroscopic' level. Then fundamental physical theory would remain fundamentally vague, until concepts like 'macroscopic' could be made sharper than they are today [16, p. 142].

Bell's theorem is generally considered as a proof of nonlocality, as in the third item of Bell's quote. But, as Bell's theorem is encompassed by Kochen-Specker theorem about contextuality, the second item of Bell's quote appears to be the most relevant, this alternative *rules out the introduction of exophysical automatons-with a random behavior-let alone observers endowed with free will. If you are willing to accept that option, then it is the entire universe which is an indivisible, nonlocal entity* [4, p. 173].

Bell's theorem consists of an inequality that is obeyed by dichotomic classical variables but is violated by the (dichotomic) eigenvalues of a set of quantum operators. The simplest form of Bell's arguments [4, p. 174] makes use of four observables σ_i, $i = 1, 2, 3, 4$, taking values in $\{-1, 1\}$, of which Bob can measure (σ_1, σ_3) and Alice (σ_2, σ_4). One introduces the number

$$C = \sigma_1\sigma_2 + \sigma_2\sigma_3 + \sigma_3\sigma_4 - \sigma_4\sigma_1 = \pm 2$$

and observes the (so-called Bell/Clauser-Horne-Shimony-Holt (CHSH)/Cirel'son's) inequality [4, p. 164]

$$|\langle\sigma_1\sigma_2\rangle + \langle\sigma_2\sigma_3\rangle + \langle\sigma_3\sigma_4\rangle - \langle\sigma_4\sigma_1\rangle| \leq 2,$$

where $\langle\rangle$ here means that we are taking averages over many experiments. This inequality holds for any dichotomic random variables σ_i that are governed by a joint probability distribution. Bell's theorem states that the aforementioned inequality is violated if one considers quantum observables with dichotomic eigenvalues. An illustrative example is the following set of two-qubit observables: $\sigma_1 = IX$, $\sigma_2 = XI$, $\sigma_3 = IZ$, and $\sigma_4 = ZI$, where X, Y and Z are the ordinary Pauli spin matrices and where, e.g., IX is a short-hand for $I \otimes X$ (used also in the sequel).

The norm $||C||$ of C (see [4] or [9] for details) is found to obey $||C|| = 2\sqrt{2} > 2$, a maximal violation of the aforementioned inequality.

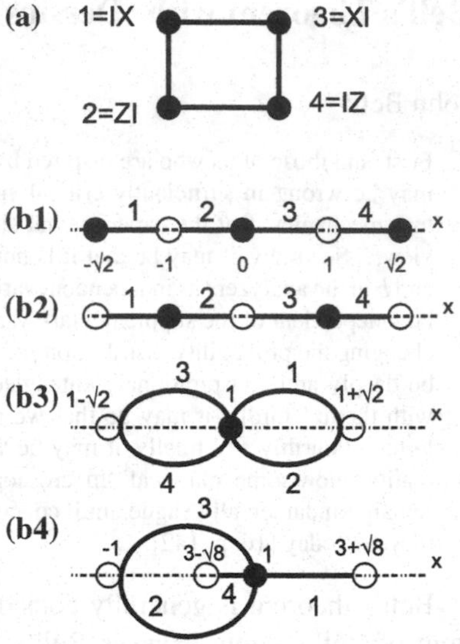

Fig. 4.2 A simple observable proof of Bell's theorem is embodied in the geometry of a (properly labeled) square (a) and four associated *dessins d'enfants*, (b_1) to (b_4). For each *dessin* an explicit labeling of its edges in terms of the four two-qubit observables is given. The (real-valued) coordinates of black and white vertices stem from the corresponding Belyi functions as explained in the main text

The point-line incidence geometry associated with our four observables is one of the simplest, that of a square—Fig.4.2a; each observable is represented by a point and two points are joined by a segment if the corresponding observables commute. It is worth mentioning here that there are altogether 90 distinct squares among two-qubit observables and as many as 30240 when three-qubit labeling is employed, each yielding a maximal violation of the Bell-CHSH inequality.

Dessins D'enfants for the Square and Their Belyi Functions

The methodology described at section "The Manifolds Traits of a 'Dessin D'enfant" was used to arrive at the result that the geometry of the square/quadrangle can be generated by four different *dessins*, (b_1), ..., (b_4), associated with permutations groups P isomorphic to the dihedral group D_4 of order 8. Two of them, (b_1) and (b_2) are tree-like and the other two, (b_3) and (b_4), contain loops.

The first *dessin* (b_1) has the signature $s = (B, W, F, g) = (3, 2, 1, 0)$ and the symmetry group $P = \langle(2, 3), (1, 2)(3, 4)\rangle$ whose cycle structure reads $[2^1 1^2, 2^2, 4^1]$, i.e. one black vertex is of degree two, two black vertices have degree one, the two white vertices have degree two and the face has degree four. The corresponding Belyi function reads $f(x) = x^2(2 - x^2)$, see [9] for details. The Belyi functions for the other cases (b_2) to (b_4) are $f(x) = (x^2 - 1)^2$, $f(x) = \frac{(x-1)^4}{4x(x-2)}$ and $f(x) = \frac{(x-1)^4}{16x^2}$, respectively. Just observe that (critical) points of the dessin, where the derivative $f'(x)$ vanishes, correspond to black points where the valency is larger than one, that black point coordinates correspond to solutions of the equation $f(x) = 0$, that white point coordinates correspond to the solution of the equation $f(x) = 1$ and that the number of loops reflects in the number of poles of the corresponding Belyi function.

It is intriguing to see that all coordinates of a dessin live in the extension field $\mathbb{Q}(\sqrt{2})$ of the rational field $\mathbb{Q}$. Hence, a better understanding of the properties of the group of automorphisms of this field may lead to fresh insights into the nature of this important theorem of quantum physics.

Kochen-Specker Theorem with "Dessins D'enfants': Two Qubits

I am grateful to N. D. Mermin for patiently explaining to me that Ref. 11 [A. Peres, *Phys. Lett. A* **151**, 107 (1990); *Found. Phys.* **22**, 357 (1992)]) was a Kochen-Specker argument, not one about locality, as I had wrongly thought [17].

Bell's theorem is a no-go theorem that forbids local hidden variable theories. Kochen-Specker theorem [5] is stronger by placing new constraints in the permissible types of hidden variable theories. Kochen-Specker theorem forbids the simultaneous validity of the two statements (i), that all hidden variables have definite values at a given time (value definiteness) and (ii), that those variables are independent of the setting used to measure them (non-contextuality). Thus, quantum observables cannot represent the 'elements of reality' of EPR paper [1].

Kochen-Specker theorem establishes that even for compatible/commuting observables A and B with values $v(A)$ and $v(B)$, the equations $v(aA + bB) = av(A) + bv(B)$ $(a, b \in \mathbb{R})$ or $v(AB) = v(A)v(B)$ may be violated. The authors restricted the observables to a special class, viz. so-called yes-no observables, having only values 0 and 1, corresponding to projection operators on the eigenvectors of certain orthogonal bases of a Hilbert space.

One of the simplest types of violation is a set of nine two-qubit operators arranged in a 3×3-grid [6]. This grid is a remarkable one: all triples of observables located in a row or a column are mutually commuting and have their product equal to $+II$ *except for* the middle column, where $XX.YY.ZZ = -II$. Mermin was the first to observe that this is a Kochen-Specker (parity) type contradiction since the product of all triples yields the matrix $-II$, while the product of corresponding eigenvalues is $+1$ (since each of the latter occurs twice, once in a row and once in a column) [7]. Note that the Mermin square comprises a set of nine elementary squares/quadrangles that themselves constitute a proof of Bell's theorem, as shown at the preceding section.

The Mermin 'magic' square may be used to provide many contextuality proofs from the vectors shared by the maximal bases corresponding to a row/column of the diagram. The simplest, a so-called (18, 9) proof, (18 vectors and 9 bases) has, remarkably, the orthogonality diagram which is itself a Mermin square (9 vertices for the bases and 18 edges for the vectors) [20, Eq. (6)].

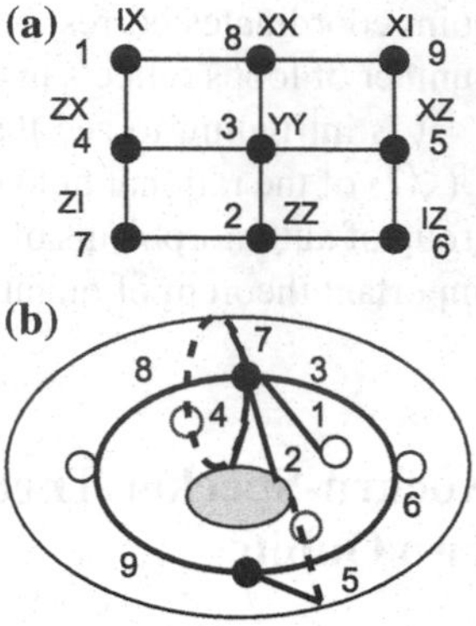

Fig. 4.3 A 3 × 3 grid with points labeled by two-qubit observables (*aka* a Mermin magic square) (**a**) and a stabilizing *dessin* drawn on a torus (**b**)

A 'Dessin D'enfant' for the Mermin Square

Mermin square is shown Fig. 4.3a. One can recover this geometry with a genus one *dessin*, with signature (2, 5, 2, 1), as shown in Fig. 4.3b. The corresponding permutation group is $P = \langle(1, 2, 4, 8, 7, 3)(5, 9, 6), (2, 5)(3, 6)(4, 7)(8, 9)\rangle \cong \mathbb{Z}_3^2 \rtimes \mathbb{Z}_2^2$, having the cycle structure $[6^1 3^1, 2^4 1^1, 6^1 3^1]$. This *dessin* lies on a Riemann surface that is a torus (not a sphere $\hat{\mathbb{C}}$), being thus represented by an elliptic curve. The topic is far more advanced and we shall not pursue it in this paper (see, e.g., [21] for details). The stabilizer of a pair of edges of the *dessin* is either the group $\mathbb{Z}_2$, yielding Mermin's square M_1 shown in Fig. 4.3a, or the group $\mathbb{Z}_1$, giving rise to a different square M_2 from the maximum sets of mutually non-collinear pairs of points of M_1. The union of M_1 and M_2 is the Hesse configuration.

Kochen-Specker Theorem with 'Dessins D'enfants': Three Qubits

Poincaré wrote:

> Perceptual space is only an image of geometric space, an image altered by a sort of perspective [22, p. 342] and Weyl wrote: In this sense the projective plane and the color continuum are isomorphic with one another [22, p. 343].

Color experience through our eyes to our mind relies on the real projective plane $\mathbb{RP}^2$ [22]. Three-qubit contextuality also relies on $\mathbb{RP}^2$ thanks to a Mermin 'magic' pentagram, that for reasons explained below in (i) we denote $\bar{\mathcal{P}}$ (by abuse of language because we are at first more interested to see the pentagram as a geometrical configuration than as a graph). One such a pentagram is displayed in Fig. 4.4a. It consists of a set of five lines, each hosting four mutually commuting operators and any two sharing a single operator. The product of operators on each of the lines is $-III$, where I is the 2×2 identity matrix. It is impossible to assign the dichotomic truth

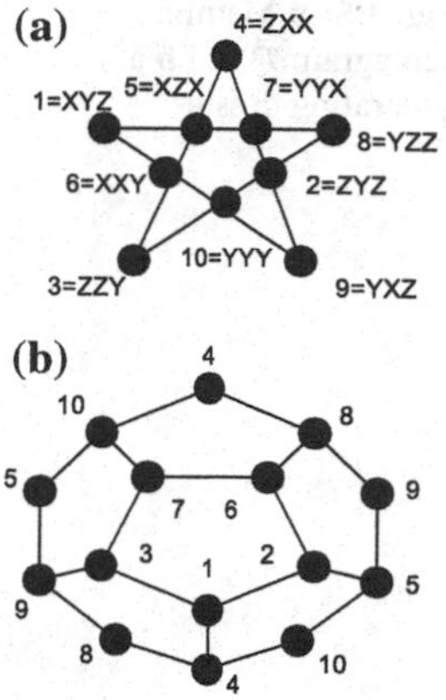

Fig. 4.4 **a** A Mermin pentagram $\bar{\mathcal{P}}$ and **b** the embedding of the associated Petersen graph $\mathcal{P}$ on the real projective plane as a hemi-dodecahedron

values ± 1 to eigenvalues while keeping the multiplicative properties of operators so that the Mermin pentagram is, like its two-qubit sibling, 'magic', and so contextual [6, 19, 20].

Let us enumerate a few remarkable facts about a pentagram.

(i) The graph $\bar{\mathcal{P}}$ of a pentagram is the complement of that of the celebrated Petersen graph $\mathcal{P}$. One noticeable property of $\mathcal{P}$ is to be the smallest bridgeless cubic graph with no three-edge-coloring. The Petersen graph is thus not planar, but it can be embedded without crossings on $\mathbb{RP}^2$ (one of the simplest non-orientable surfaces), as illustrated in Fig. 4.4b.

(ii) The Petersen graph may also be seens as the complement of the Desargues configuration 10_3 (a celebrated projective geometry that has ten lines with three points and ten points each of them incident with three lines), see [9, Fig.11].

(iii) There exist altogether 12096 three-qubit Mermin pentagrams, this number being identical to that of automorphisms of the smallest split Cayley hexagon *GH*(2,2)—a remarkable configuration of 63 points and 63 lines [19] pictured in Fig. 4.7a.

(iv) Now comes an item close to the *it from bit* perspective. The Shannon capacity of a graph is the maximum number of k-letter messages than can be sent through a channel without a risk of confusion. The Shannon capacity of $\mathcal{P}$ is found to be optimal and equal to 4, much larger than that $\sqrt{5}$ of an ordinary pentagon.

(v) Finally, the pentagram configuration in Fig. 4.5a may be generated/stabilized by a 'dessin d'enfant' on the Riemann sphere, having permutation group isomorphic to the alternating group A_5 and cycle structure $[3^2 1^1, 2^4 1^2, 5^2]$, as shown in Fig. 4.5b. The stabilizer of a pair of edges of the *dessin* is either the group $\mathbb{Z}_1$ giving rise to the Mermin's pentagram, or the group $\mathbb{Z}_2$ giving rise to the Petersen graph.

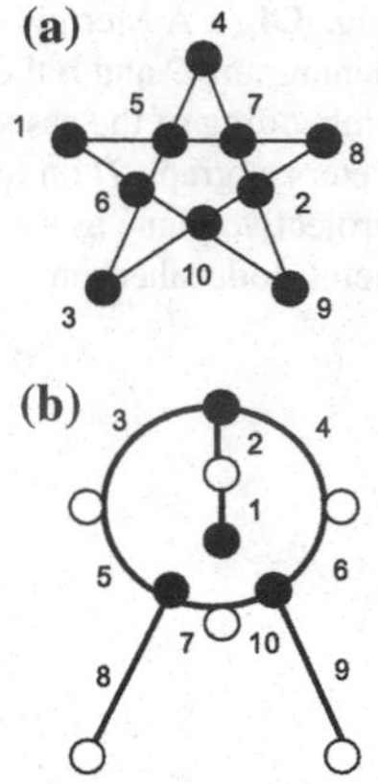

Fig. 4.5 **a** Mermin pentagram $\bar{\mathcal{P}}$ and **b** a generating dessin

Dessins D'enfants and Generalized Polygons

Jacques Tits:

> I would say that mathematics coming from physics is of high quality. Some of the best results we have in mathematics have been discovered by physicists. I am less sure about sociology and human science [23].

Jacques Tits discovered generalized polygons (also called generalized n-gons). A generalized polygon is an incidence structure between a discrete set of points and lines whose incidence graph has diameter n (the maximum eccentricity of any vertex) and girth $2n$ (the length of a shortest cycle). A generalized polygon of order (s, t) has every line containing $s + 1$ points and ever point lying on $(t + 1)$ lines. Remarkably, the generalized 4-gon/quadrangle of order (2, 2), namely $GQ(2, 2)$ controls the commutation structure of the 15 two-qubit observables [18] and the generalized 6-gon/hexagon $GH(2, 2)$ does the job for the 63 three-qubit observables [19].

An important concept pertaining to generalized polygons is that of a geometric hyperplane. A geometric hyperplane of a generalized polygon is a proper subspace meeting each line at a unique point or containing the whole line. The substructure of a polygon of order $(2, t)$ highly relies on its hyperplanes in the sense that one 'adds' any two of them to form another geometric hyperplane. The 'addition' law in question is nothing but 'the complement of the symmetric difference' of the two sets of points involved in the pair of selected hyperplanes. There exists three kinds of hyperplanes in $GQ(2, 2)$ one of them being the Mermin square described in section "Kochen-Specker Theorem with 'Dessins D'enfants': Two Qubits". The structure of hyperplanes of $GH(2, 2)$ is of utmost importance to describe the three-qubit Kochen-Specker theorem as described exhaustively in [19].

Next, we find that generalized polygons are induced/stabilized by dessins d'enfants. As for the case of the geometry of the square, a selected geometry $\mathcal{G}$ may be induced/stabilized by many dessins $\mathcal{D}_i$, i.e. the correspondence $f : \mathcal{D}_i \to \mathcal{G}$

is non injective. Moreover, the number of dessins grows exponentially with the number of their edges so that the systematic search of all maps f may become tedious. To simplify the search of a solution f inducing a selected $\mathcal{G}$ (such as a generalized polygon) one restricts the search to subgroups of the cartographic group C_2^+ (go back to section "The Manifolds Traits of a 'Dessin D'enfant'" for the definition).

A Dessin D'enfant for the Generalized Quadrangle GQ(2, 2)

The generalized quadrangle $GQ(2, 2)$ encodes the commutation relations of two-qubit operators as shown in Fig. 4.6a [18].

A dessin stabilizing the generalized quadrangle $GQ(2, 2)$ may be obtained by studying the conjugacy classes of the subgroup $C_2^+/[\rho_2^4 = 1]$, whose permutation representation of the cosets is isomorphic to the symmetry group of $GQ(2, 2)$, that is the symmetric group S_6. This is shown in Fig. 4.6. One finds that the dessin in Fig. 4.6b has two types of stabilizers for a pair od edges, one isomorphic to $\mathbb{Z}_2^5$ and inducing $GQ(2, 2)$ and the other one isomorphic to $\mathbb{Z}_6$ and inducing the complement of $GQ(2, 2)$.

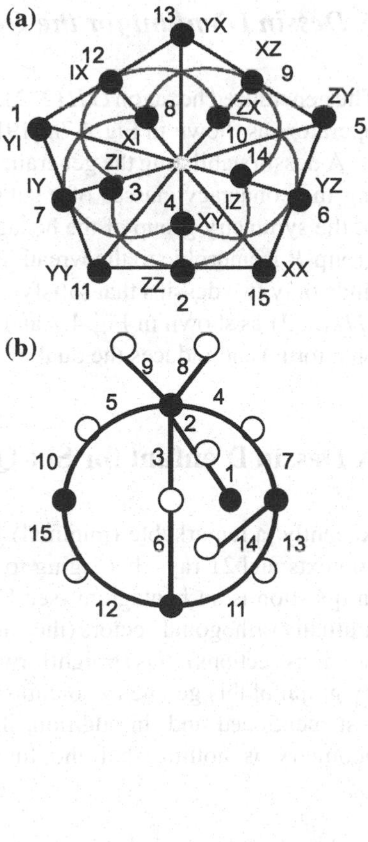

Fig. 4.6 The generalized quadrangle $GQ(2, 2)$ **(a)** with its points labeled by the elements of the two-qubit Pauli group and a stabilizing *dessin* **(b)**

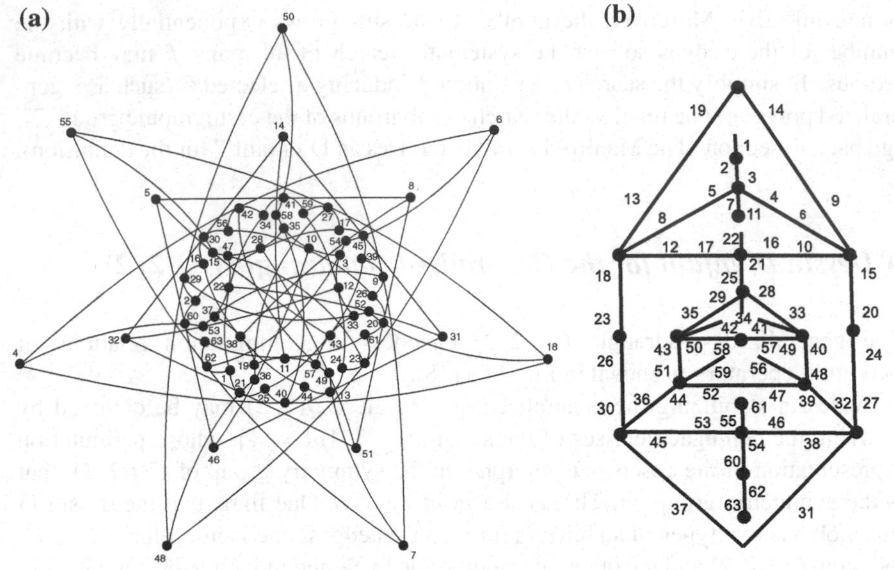

Fig. 4.7 **a** The hexagon $GH(2, 2)$ and **b** the *dessin* for its collinearity graph. To simplify the drawing, white points are not shown but half-edges are labelled

A Dessin D'enfant for the Generalized Hexagon GH(2, 2)

The generalized hexagon $GH(2, 2)$ encodes the commutation relations of three-qubit operators as shown in Fig. 4.7a [19].

A dessin stabilizing the generalized hexagon $GH(2, 2)$ may be obtained by studying the conjugacy classes of a subgroup of C_2^+ whose finite representation is that of the symmetry group of the hexagon. Then one selects the dessins of permutation group P isomorphic to the wreath product $S_3 \wr S_3$ (of order 1296). Remarkably, one finds only two dessins that satisfy these requirements, one is of genus 0 and induces $GH(2, 2)$ as shown in Fig. 4.7 and the other one (not shown) is of genus 1 (drawn on a torus) and induces the dual of $GH(2, 2)$.

A Dessin D'enfant for Six-Qudit Contextuality

Recently, a remarkable (minimal) Kochen-Specker configuration built from seven contexts and 21 rays, belonging to a six-qudit system, has been built [14]. The set in question is an heptagram (see Fig. 4.8a) in which the seven lines are hexads of mutually orthogonal vectors (they are not quantum observables as was the case at the previous sections). It is straightforward to check (with a computer) that the collinearity graph of this geometry contains two kinds of maximal cliques, the seven hexads just mentioned and, in addition, 35 triangles. The graph of the latter 35-triangle geometry is nothing but the line graph of the complete graph K_7 [the line

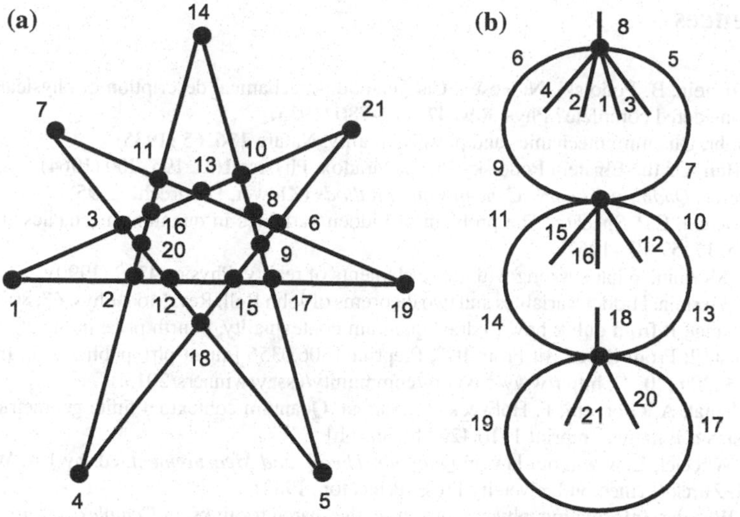

Fig. 4.8 **a** The seven-context geometry of six-qudit contextuality and **b** a dessin stabilizing the 35-triangle geometry lying in (**a**). To simplify the drawing, white points are not shown but half-edges are labelled

graph of the complete graph K_5 is the Desargues configuration that occured at section "Kochen-Specker Theorem with "Dessins D'enfants' Two Qubits", item (ii)]. The symmetry group of the 35-triangle geometry, as that of the heptagram, is the seven-letter symmetric group S_7. Starting from a finite representation of S_7 that underlies a subgroup of the cartographic group C_2^+, it is not difficult to find a dessin stabilizing the aforementioned 35-triangle geometry, as shown in Fig. 4.8b. Once again, it has been shown that quantum contexts are intimately related to dessins d'enfants.

Conclusion

It was not anticipated by his creator that geometries induced by 'dessins d'enfants' would so nicely fit the drawings/geometries underlying quantum contextuality. I believe that this key observation opens new vistas for the interpretation of quantum measurements in terms of algebraic curves over the rationals. The symmetries of general dessins rely on a fascinating group called the universal (or absolute) Galois group over the rationals, a still rather mysterious object. May be the present work gives some substance to an hidden variable interpretation of the quantum world, as hoped by Einstein, but in a subtle way. We just mentioned in passing (in section "Dessins D'enfants and Generalized Polygons") that the geometries relevant to quantum contexts have a rich substructure of hyperplanes that also has to be incorporated in the new design. More details will be given at a next stage of our research.

References

1. A. Einstein, B. Podolski, N. Rosen, Can quantum-mechanical description of physical reality be considered complete? Phys. Rev. **47**, 777–780 (1935)
2. N. Bohr, Quantum mechanics and physical reality. Nature **136**, 65 (1935)
3. J.S. Bell, On the Einstein Podolsky Rosen paradox. Physics **1**(3), 195–200 (1964)
4. A. Peres, *Quantum Theory: Concepts and Methods* (Kluwer, Dordrecht, 1995)
5. S. Kochen, E.P. Specker, The problem of hidden variables in quantum mechanics. J. Math. Mech. **17**, 59–87 (1967)
6. N.D. Mermin, What's wrong with these elements of reality. Physics **43**, 9 (1990)
7. N.D. Mermin, Hidden variables and two theorems of John Bell. Rev. Mod. Phys. **65**, 803 (1993)
8. M. Planat, It from qubit: how to draw quantum contextuality, Fourth price in the 2013 FQXi contest "It From Bit or Bit From It?", Preprint 1306.0356 [quant-ph], published in Information **5**, 209 (2014). http://www.fqxi.org/community/essay/winners/2013.1
9. M. Planat, A. Giorgetti, F. Holweck, M. Saniga, Quantum contextual finite geometries from Dessins d'Enfants, Preprint 1310.4267 [quant-ph]
10. J.A. Wheeler, Law without law, in *Quantum Theory and Measurement*, ed. by J.A. Wheeler, W.H. Zurek (Princeton University Press, Princeton, 1983)
11. J.A. Wheeler, Information, physics, quantum: the search for links, in *Complexity, Entropy, and the Physics of Information*, ed. by W. Zurek (Addison-Wesley, Redwood City, 1990)
12. A. Grothendieck, Sketch of a programme, written in 1984 and reprinted with translation, in *Geometric Galois Actions 1. Around Grothendieck's Esquisse d'un Programme, 2. The Inverse Galois Problem, Moduli Spaces and Mapping Class Groups*, ed. by L. Schneps, P. Lochak (Cambridge University Press, Cambridge, 1997)
13. L. Schneps (ed.), *The Grothendieck Theory of Dessins d'Enfants* (Cambridge University Press, Cambridge, 1994)
14. P. Lisonek, P. Badziag, J.R. Portillo, A. Cabello, Kochen-Specker theorem with seven contexts. Phys. Rev. A **89**, 042101 (2014)
15. S.K. Lando, A.K. Zvonkin, *Graphs on Surfaces and Their Applications* (Springer, Berlin, 2004)
16. J. Bell, Bertlmann's socks and the nature or reality. J. Phys. Colloque **42**, C2-41–C2-62 (1981); reprinted in *John S. Bell On the Foundations of Quantum Mechanics* (M. Bell et al. eds, World Scientific, Singapore, 2001), pp. 126–147
17. A. Peres, Generalized Kochen-specker theorem. Found. Phys. **26**, 807 (1996)
18. M. Saniga, M. Planat, Multiple qubits as symplectic polar spaces of order two. Adv. Stud. Theor. Phys. **1**, 1–4 (2007)
19. M. Planat, M. Saniga, F. Holweck, Distinguished three-qubit 'magicity' via automorphisms of the split Cayley hexagon. Quant. Inf. Process. **12**, 2535 (2013)
20. M. Planat, On small proofs of the Bell-Kochen-Specker theorem for two, three and four qubits. EPJ Plus **127**, 86 (2012)
21. E. Girondo, G. González-Diez, *Introduction to Riemann Surfaces and Dessins d'Enfants*, The London Mathematical Society (Cambridge University Press, Cambridge, 2012)
22. B. Flanagan, Are perceptual fields quantum fields? Neuroquantology **3**, 334 (2003)
23. M. Raussen, C. Skau, Interview with John G. Thompson and Jacques Tits. Not. AMS **56**, 478 (2009)

Chapter 5
The Tao of It and Bit

Ovidiu Cristinel Stoica

Fourth prize in the FQXi's 2013 Essay Contest 'It from Bit, or Bit from It?'.

—To J.A. Wheeler, at 5 years after his death.

Abstract The main mystery of quantum mechanics is contained in Wheeler's delayed choice experiment, which shows that the past is determined by our choice of what quantum property to observe. This gives the observer a *participatory role* in deciding the past history of the universe. Wheeler extended this participatory role to the emergence of the physical laws (*law without law*). Since what we know about the universe comes in yes/no answers to our interrogations, this led him to the idea of *it from bit* (which includes the participatory role of the observer as a key component). The yes/no answers to our observations (*bit*) should always be compatible with the existence of at least a possible reality—a global solution (*it*) of the Schrödinger equation. I argue that there is in fact an interplay between *it* and *bit*. The requirement of *global consistency* leads to apparently acausal and nonlocal behavior, explaining the weirdness of quantum phenomena. As an interpretation of Wheeler's *it from bit* and *law without law*, I discuss the possibility that the universe is mathematical, and that there is a "mother of all possible worlds"—named the *Axiom Zero*.

Wheeler

John Archibald Wheeler was, arguably, the most influential physicist since Einstein, contributing to radical insights in general relativity, quantum mechanics, quantum field theory, quantum gravity, to mention just a few domains. Much of this influence

O.C. Stoica (✉)
Department of Theoretical Physics, Horia Hulubei National Institute for Physics and Nuclear Engineering, Str. Reactorului no.30, MG-6, Bucharest - Magurele, Romania
e-mail: crististoica@theory.nipne.ro

A. Aguirre et al. (eds.), *It From Bit or Bit From It?*,
The Frontiers Collection, DOI 10.1007/978-3-319-12946-4_5

was done through his many brilliant PhD students.[1] Although I've never met him, I see him as a person who is willing to risk his reputation by allowing him and his students to develop ideas which apparently contradicted the very foundations of physics, as accepted in his time. He worked on radical (at least for that time) subjects like *wormholes*, *black holes*, *geons* (objects made just of spacetime, including a way to obtain the mass and the electromagnetic field as effects of the topology of spacetime [1]), *wavefunction of the universe*, with the accompanying *end of time*, strange superpositions of different topologies in a *quantum foam*, *delayed choice experiments* which seem to imply that the observer affects the past [2–4]. Moreover, the initial conditions of the observed system have to depend on those of the measurement device [5].

He encouraged his students to challenge well established paradigms, with ideas like:

- A particle goes from one point to another by following all possible paths, even if it goes faster than light, or even back in time [6].
- Many features of Quantum Mechanics can be better understood if we admit that there are many worlds [7, 8].
- Black holes have their own thermodynamics, including entropy [9]. When combining the effect discovered by another student of Wheeler, Bill Unruh, with the principle of equivalence, we obtain the Hawking(-Zel'dovich-Starobinski) radiation.

Imagine how a PhD student coming with one of the above-mentioned ideas would be perceived. Such theories, even nowadays, appear to many as taken from science fiction, if not from new-age pseudoscience. How such ideas, instead of being ridiculed, were even accepted as top science? I would thank Wheeler's courage for the new generation of Einsteins who appeared and changed the face of modern physics—if one genuinely wants to find or foster new Einsteins [10], one has so much to learn from him. And when his former students became widely acknowledged, he modestly remained in the shadow.

These beautiful theories were well-developed, to derive qualitative and quantitative predictions. Many of them were experimentally confirmed, while others are still waiting, and some just stand as beautiful concepts, whose role is to explain phenomena, rather than predicting new ones. Some of his ideas are so visionary, that probably we will never be able to verify them completely by experiments. His proposal *it form bit* [2, 11–14] combines in an amazing way his previous results, and those of his students. This makes the subject of this essay.

[1] From Wheeler's students, I will mention only a few who changed the face of physics: Richard Feynman, Hugh Everett III, Jacob Bekenstein, Warner Miller, Robert Geroch, Charles Misner, Kip Thorne, Arthur Wightman, Bill Unruh, Robert Wald, Demetrios Christodoulou, Ignazio Ciufolini, Kenneth Ford, and others, to whom I apologize for not mentioning.

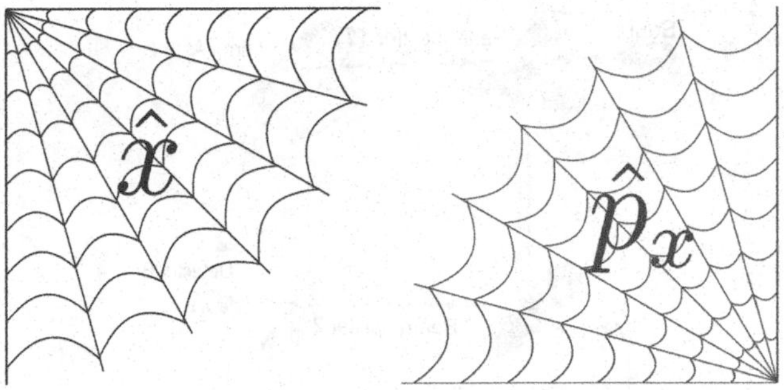

Fig. 5.1 Spiderweb for catching particles, and spiderweb for catching waves

It-sy Bit-sy Spider

Imagine a world in which there are three kinds of beings: spiders, flies, and dragonflies. Spiders eat flies and dragonflies, but unfortunately they can't fly, so to catch their food, they have to build webs. Imagine there are two kinds of webs, one kind can catch only flies, and the other one can catch only dragonflies. So far nothing weird.

Now imagine spiders can see the prey flying, but their sight is not as good to detect what kind the prey is. They only see that whenever an insect flies towards a web, it is caught (Fig. 5.1).

Spiders are very intrigued, because they wonder:

> What we catch in a web-for-flies, is always a fly. What if we replace in the last moment the web-for-flies with a web-for-dragonflies? Obviously, in this case we would catch a dragonfly. But how can the kind of the prey be decided by our choice of the web? Was the prey a fly, or a dragonfly, before being caught in the web?

A quantum world is similar to a world in which the spider's choice of the type of the web determines what species is the insect which already flies toward the web.

Wheeler's *delayed choice experiment* can be seen as switching in the last moment the web with another kind of web, while the insect is still heading toward the web.

Delayed Choice Experiment

Recall the quantum experiment based on the Mach-Zehnder interferometer. Light is emitted by a source, and split by beam splitter 1 (see Fig. 5.2). The two halves of the ray are redirected by two mirrors to meet again, and the original ray is recomposed, by beam splitter 2. The photons always trigger detector B.

Now, remove the beam splitter 2. The photons will trigger with equal probability both detectors A and B (Fig. 5.3).

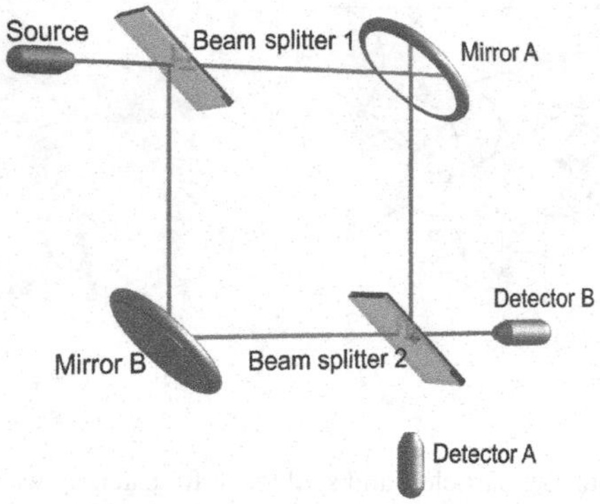

Fig. 5.2 *Both ways* observation

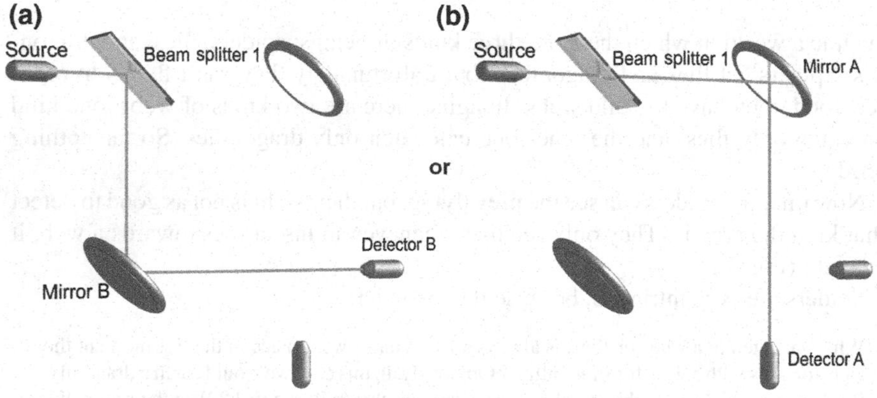

Fig. 5.3 *Which-way* observation

Wheeler proposes to delay the decision of whether to keep or to remove the beam splitter 2, until we are sure the photon passed from splitter 1 [2]. In fact, his thought experiment uses instead of beam splitters and mirrors, the deflection of light caused by the gravity of an entire galaxy. He concludes [14]:

> Since we make our decision whether to measure the interference from the two paths or to determine which path was followed a billion or so years after the photon started its journey, we must conclude that our very act of measurement not only revealed the nature of the photon's history on its way to us, but in some sense *determined* that history. The past history of the universe has no more validity than is assigned by the measurements we make–now!

The delayed choice experiment is the source for Wheeler's *law without law* and *it from bit*.

Law Without Law

Wheeler pushed to the extreme his idea of delayed choice experiment. He thought that the observer determines not only the past of a quantum system, but the very physical laws! We can say that he extended his condensed formulation of Bohr's vision on quantum mechanics, "no phenomenon is a phenomenon, until it is an observed phenomenon", to "no fundamental law is a fundamental law, until it is an observed fundamental law".

Wheeler thought that the observer participates in choosing now the physical laws for the entire past and future history. He coined this vision *law without law*. He wrote in [15]

> If the views that we are exploring here are correct, one principle, observer-participancy, suffices to build everything. [...] [The picture of the participatory universe] has no other than a higgledy-piggledy way to build law: out of the statistics of billions upon billions of acts of observer-participancy each of which by itself partakes of utter randomness.

If Wheeler was right that we decide the physical laws, by our very choices as observers of the universe, then, due to their important and bold contributions to physics, he and his students are responsible for many preposterous features of our universe.

Evolving Laws

Regarding *law without law*, one may wonder how could there be different sets of laws to choose from. One possibility is that some fundamental constants are not really constants. They may became constant moments after the big-bang, frozen by symmetry breaking. Initially Wheeler proposed that after the big-crunch there will be a new universe, with different constants, but now we know that there will be no big-crunch [16]. A more recent proposal was made by Smolin, that the laws evolve from universe to *baby-universe* [17–19]. Presumably, a baby universe appears by going beyond a future spacelike singularity (like that of Schwarzschild). Penrose claims that this can't be done, because we can't match together a black hole and big-bang singularity, since they are of different types. They appear to be different, but there is an appropriate (singular) coordinate transformation which makes the Schwarzschild coordinate of the same type as the Friedmann-Lemaître-Robertson-Walker (FLRW) one [20]. In fact, at least in the case of the Oppenheimer-Snyder model of a black hole, the star is modeled as a time-reversed pure dust FLRW solution, so it is not justified to claim that the two can't be matched together (a FLRW singularity is the continuation of a time-reversed FLRW singularity [21, 22]).

But Wheeler's philosophy *law without law* goes far beyond the idea of a mechanism of random mutations of the constants. He viewed the law as being created, or perhaps chosen from an infinity of alternatives, by the very observation process. The *bit* not only determines the (past) *it* of the universe, but also the laws.

Tegmark's Mathematical Universe

Tegmark's *mathematical universe* [23, 24] can provide an implementation of Wheeler's *law without law*. Tegmark proposes that all possible mathematical structures exist, and our universe is one of them.

He said that, in order for a universe to exist, it is enough to have a simulation of it, and that it is not even needed to run the simulation, merely having the description as a string of bits written on a CD-ROM is enough.

But the meaning of a string of bits depends on the language used to encode the information in it. The first comment I want to make about this is that the meaning of the string specifying our universe can be anything, including the specifications of any other possible universe, because for any possible meaning, one can always imagine a language in which the string has that meaning. Hence, any string, for example "0", is enough to specify all possible universes, given the appropriate decoding language.

The second comment is that the language has to be specified as well, in another language, and we arrive at an infinite regress. To avoid the regress, one can admit that there is a reality given by those specifications. Perhaps also an observer is needed, a "ghost in the quantum Turing machine" [25], something that "breathes fire into the equations" [26].

It from Bit

Wheeler tried to remove completely the idea of an independent reality (*it*), proposing that it emerges from the information contained in our observations (*bit*), which is the only one existent [14]:

> it is not unreasonable to imagine that information sits at the core of physics, just as it sits at the core of a computer

and

> I build only a little on the structure of Bohr's thinking when I suggest that we may never understand this strange thing, the quantum, until we understand how information may underlie reality. Information may not be just what we *learn* about the world. It may be what *makes* the world.

Wheeler's *it from bit* claims that the information is fundamental, more fundamental than anything else. But it is not simply a *digital theory of everything*. The central point is indeed the bit, the information about the universe, which is accessible to the observer. But equally important is the fact that the observer has a participatory role.

Wheeler often represented the universe as the letter 𝕌, with the big-bang at the right end of the curve which makes the letter, and the observer at the left end, represented as an eye which, by mere observation, brings into existence the entire past history of the universe.

This is why Wheeler's *it from bit* should not be used to support the version of digital physics which just claims that "everything is information"; nor should it be rejected by reducing his ideas to that idea [27]. Wheeler made a much more profound point than that, as we have seen.

On the other hand, most of his arguments are based on the fact that we can only know bits of information, and on the delayed choice experiment. Besides the participatory role of the observer, which is difficult to deny, one should admit that the bits are subjective, pertaining to the observer. The fact that we can only collect bits of information doesn't really mean that there is nothing else but information.

Can the Clicks of the Detectors Provide a Complete Description of Nature?

Is it possible to obtain *it* just from *bit*?

It is true that all quantum phenomena, no matter how weird they appear, are predicted by the very postulates of quantum mechanics. Strange behaviors such as correlations between the outcomes of measurements separated in space, and the fact that they depend on the context of the measurement, all follow from the simple postulates of quantum mechanics. Many try to find a more intuitive explanation for these phenomena, but they are simply explained by the fact that one can't simultaneously observe all properties of particles, because these properties are not well defined simultaneously [28].

While it is undeniable that quantum mechanics is so successful, can we know everything about the universe just by quantum measurements? Can we even guess the physical laws from the outcomes of these measurements?

There is a big obstacle which prevents us for doing this. According to the postulates of quantum mechanics, the state of a system is represented by a vector in a complex vector space (the *state space* or the *Hilbert space*). But in a vector space there is no preferred basis, and the postulates of quantum mechanics are independent of any such basis. In reality, we know that the vector space containing all possible states has a richer structure, that the position in the physical space provides a preferred basis. We also know that each type of particle comes with its own state space, and the total space is obtained by taking tensor products between copies of these one-particle spaces. But these can't result simply by looking at the outcomes of quantum measurements, because the same outcomes would be obtained if the state vector of the universe is rotated in the state space by a unitary transformation. This shows that the information about the position basis and the tensor product structure is not encoded in the outcomes of measurements, so *it* doesn't simply follow from *bit* [29].

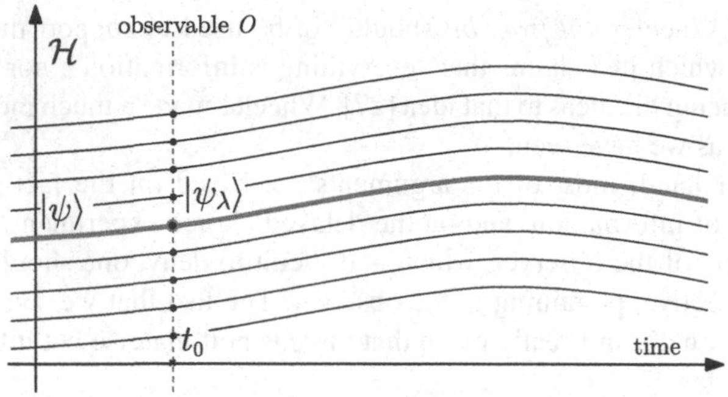

Fig. 5.4 Any property we choose to observe, it is well-defined only for a small subset of the possible states. The observed system turns out to be in such a state

Delayed Initial Conditions

As a metaphor for the participatory universe, Wheeler mentions the game of twenty questions—the player has to determine a word, by asking yes/no questions. The twist is that the word is not chosen at the beginning, but as the player asks the questions.

To make this work, the respondents have to take care that their combined answers still define a real word. This can only be done if they maintain, explicitly or not, a list of possible words. But it has to be at least one word on the list, at any moment.

What does this tell us about the universe? Classically, the state of the universe at any moment of time is determined by the initial conditions. This is prohibited in quantum mechanics, because we can only ask whether the system is in a small subset of possible states—those particular states for which the property we measure is well-defined (Fig. 5.4). It is not possible, even in principle, to know the complete state. There are no universal spiderwebs: each spiderweb can catch either flies, or dragonflies.

The observer asks questions, and the universe gives yes/no answers – bits. But the answers always define at least a possible solution.[2] It is not like there is no solution at all, as the catch-phrase *it from bit* implies. Hence, one cannot infer that nothing exists, except the outcomes of the measurements. Rather, that at any given moment of time, there are possible realities which are compatible with those answers.

This is why I think that the complete picture is not *it from bit*, but rather *it from bit & bit from it*. The yes/no questions select a subset among the possible solutions of the Schrödinger equation, but the possible answers to the yes/no questions are determined by the possible solutions which remained (Fig. 5.5) [30–32].

In addition, delayed initial conditions provide a way that free-will is compatible with deterministic laws [25, 31–33].

[2] If the initial conditions are fully specified, the solution is unique. But our observations allow us to specify only partially the initial conditions, and that's why there are more possible solutions.

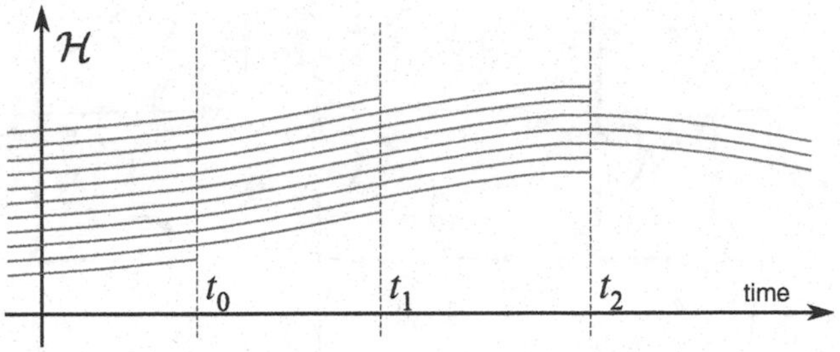

Fig. 5.5 Delayed initial conditions select possible realities, even in the past

Global Consistency Principle

Just because we don't have access to reality, but only to the bits, it doesn't mean that there is no reality. Which possibility is simpler: (1) that the yes/no bits are consistent with one another, that the probabilities are correlated, and that's all, or (2) that at any moment there is at least one possible reality, which ensure the consistency and the correlations? Isn't simpler and more logical the idea that *it* is something that prevents *bits* from contradicting one another, a "reality check".

Think at the way Schrödinger derived the energy levels from his equation. He had the equation, but he obtained the energy levels only after throwing away the solutions with bad behavior at infinity. The remaining solutions have, for an electron in an atom, a discrete spectrum. This provided the correct account to de Broglie's insight, that the wavelength of the electron's wave fits an integral number of times in the orbit. A global condition—the boundary condition at infinity—led him to the selection of only a discrete subset of solutions from the continuum set of possible solutions of Schrödinger's equation (Fig. 5.6).

But how can the solution near an atom know how to be, so that it behaves well at infinity? This is a key question. If we think in terms of disparate bits, this can't hold in a natural way. If we think that the physical solutions have reality, it becomes natural to admit that they have to behave well at infinity (otherwise they can't have physical reality).

The *global consistency principle* generalizes the boundary conditions idea, and requires that no matter how are the observations spread in spacetime, there has to be a real solution for which the observations give the observed outcomes. For example,

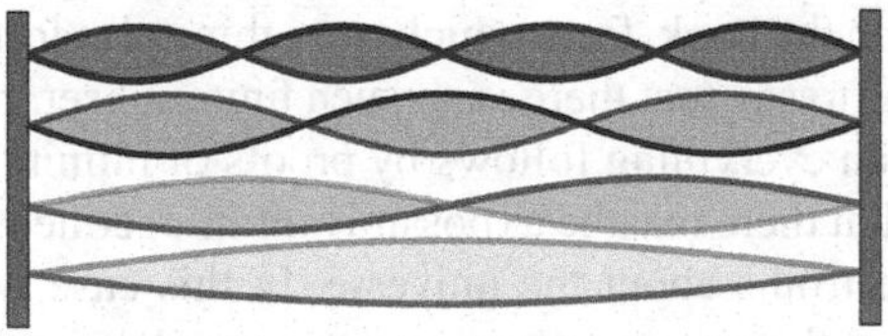

Fig. 5.6 The role of global conditions

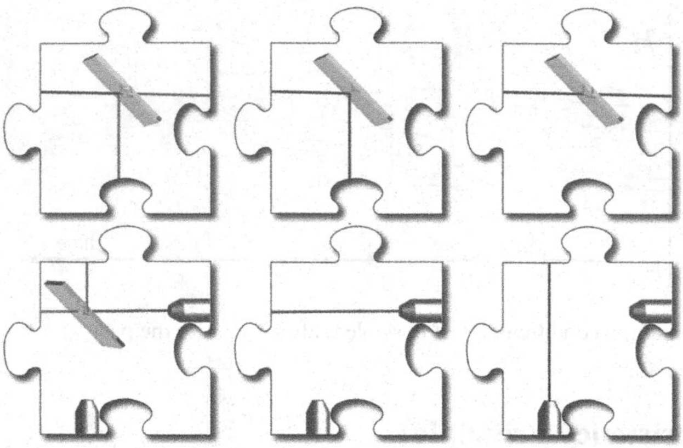

Fig. 5.7 *Global consistency principle* requires that it has to exist a solution (*it*) which combines consistently all the pieces of the puzzle (the yes/no *bits* at different points and moments of time)

it requires that the presence or absence of the beam splitter 2 in the experiment with the Mach-Zehnder interferometer has to be correlated with what happened with the ray at the beam splitter 1 (Fig. 5.7).

To understand global consistency, it may help to remember that the solutions are defined on a four-dimensional spacetime, and to think in terms of an out-of-time view, like the block universe.

The Big Book of the Universe

Here is why I find compelling the idea that our universe is mathematical. First, what we learn about anything, are *relations*. We don't know what water is, but we know its relation to our senses. Even its physical and chemical properties, follow in fact from interactions, hence from relations. Everything we know is defined by its relation with something else. If there is anything that can be mathematized, this is the *relation*. In fact, any mathematical structure is a set, along with a collection of relations defined between that set and itself [34] (Fig. 5.8).

Second, let's say that there is *a book containing every truth*. It will therefore contain the physical laws, and any truth about the state of a system at a given time—the full description of the universe. Possibly the book is infinite. Maybe there is a finite subset of propositions in the book, from which everything else follows, or maybe not. Gödel's theorem seems to say that there is no such finite subset, but maybe there is a finite subset from which everything follows by proofs of infinite length. Anyway, it seems very plausible that there may be a (possibly infinite) collection of propositions which contains all the truths about the universe. In this case, we have a a theory (of everything). To the theory we can associate a model, in the sense of *model*

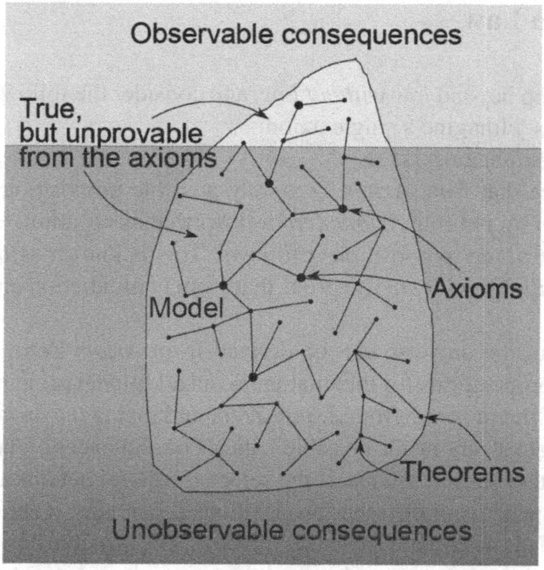

Fig. 5.8 The iceberg represents the *mathematical model* of the physical world. Its points represent true propositions, most of them unprovable from the axioms. The *large dots* represent *axioms*, and the *small dots* consequences derived from them, or *theorems*. The tip of the iceberg is what we can test by experiments and observations, at least in principle—these are the *observable consequences*. The largest part of the iceberg consists in untestable, or *unobservable consequences*

theory [35]. A mathematical model is just a set with a collection of relations between its elements, a mathematical structure. So, whatever the collection of the truths about the universe is, the same propositions hold for that mathematical structure. The universe is isomorphic to a mathematical structure [36].

"Wait!", one may say, "how about love, music, God, and so on? Are you claiming that these are just parts of a mathematical structure?" Well, so long as these concepts are confined to a set of propositions, they are isomorphic to a mathematical structure. But what is wrong with this? For many, mathematics IS love, music, God... Maybe they have the "fine ear" for mathematics, maybe they hear in it the "music of the spheres" more than others, just like some have the "fine ear" for music.

Anyway, if one believes there are things that are not included in the mathematical model of the universe, one should describe those things. And this means that one has to build propositions about them, and to describe their relations with other things. And this means that they are already present in the book of all true propositions, and implicitly in the mathematical model.

From Chaos to Law

One can go one step beyond *law without law*, and consider the following "mother of all possible worlds". Imagine a single axiom:

Axiom Zero. *Axiom Zero* is false.

It is easy to see that from *Axiom Zero*, any possible proposition follows. Let's denote *Axiom Zero* by p. From *Axiom Zero* follows that its negation, $\neg p$, is also true. But from p and $\neg p$, any proposition q follows. This is known as *the principle of explosion*, or *ex falso quodlibet*. The proof that from contradiction anything follows is very simple.[3]

Any truth about the universe can be derived from *Axiom Zero*. Like any false and undecidable propositions for that matter. So an additional principle is needed to derive the laws of the universe from *Axiom Zero*, and that is *the principle of logical consistency*. We select, among the possible logical consequences of *Axiom Zero*, only a logically consistent subset. That is, if the selected subset contains a proposition q, or if q can be deduced from the other propositions it contains, it should not contain also its negation. This describes a possible universe. Any possible universe, including ours, can be obtained from *Axiom Zero* and the *principle of logical consistency*. So we may say that *Axiom Zero* is the "mother of all possible worlds", from which, effortlessly, any possible world appears, due to the *principle of logical consistency*.

But the *principle of logical consistency* does not tell what the laws are. We learn about the laws only by our observations, and, as Wheeler said, our observations can decide what the laws are. The outcome of each new observation is constrained to be consistent with the previous ones, so that the *principle of logical consistency* is not violated.

We arrive again at the conclusion that, to have *bits* which don't contradict one another, an underlying *it* which satisfies to those bits should exist.

[3] Assuming both propositions p and $\neg p$ are true, we want to prove q. Since p is true, $p \vee q$ is true. But since $\neg p$ is true, p is false. From $p \vee q$ and $\neg p$ follows that q is true.

References

1. C.W. Misner, J.A. Wheeler, Classical physics as geometry: gravitation, electromagnetism, unquantized charge, and mass as properties of curved empty space. Ann. Phys. **2**, 525–603 (1957)
2. J.A. Wheeler, The 'Past' and the 'Delayed-Choice' experiment, in *Mathematical Foundations of Quantum Theory*, ed. by A.R. Marlow (Academic, New York, 1978), p. 30
3. V. Jacques, E. Wu, F. Grosshans, F. Treussart, P. Grangier, A. Aspect, J.F. Roch, Experimental realization of wheeler delayed-choice Gedanken experiment. Science **315**(5814), 966–968 (2007)
4. R. Ionicioiu, D.R. Terno, Proposal for a quantum delayed-choice experiment. Phys. Rev. Lett. **107**(23), 230406 (2011)
5. O.C. Stoica, Quantum measurement and initial conditions. Preprint arXiv:quantph/1212.2601 December (2012)
6. R.P. Feynman, *QED: The Strange Theory of Light and Matter* (Princeton University Press, Princeton, 1985)
7. H. Everett, "Relative State" formulation of quantum mechanics. Rev. Mod. Phys. **29**(3), 454–462 (1957)
8. H. Everett, The theory of the universal wave function, in *The Many-Worlds Hypothesis of Quantum Mechanics*, (Press, 1973), pp. 3–137
9. J.D. Bekenstein, Black holes and the second law. Lett. Al Nuovo Cim. **4**(15), 737–740 (1972). (1971–1985)
10. L. Smolin, Why no "new Einstein"? Phys. Today **58**(6), 56–57 (2005)
11. J.A. Wheeler, "On recognizing 'law without law'", Oersted Medal response at the joint APS-AAPT meeting, New York, 25 January 1983. Am. J. Phys. **51**, 398 (1983)
12. J.A. Wheeler, World as system self-synthesized by quantum networking. IBM J. Res. Dev. **32**(1), 4–15 (1988)
13. J.A. Wheeler, Information, physics, quantum: the search for links, in *Complexity, entropy and the physics of information: the Proceedings of the 1988 Workshop on complexity, entropy, and the physics of information held 29 May–10 June, 1989 in Santa Fe, New Mexico, volume 8.* (Westview Press, 1990)
14. J.A. Wheeler, K.W. Ford, Geons, Black Holes, and Quantum Foam: A Life in Physics (W.W. Norton and Company, 1998)
15. J.A. Wheeler, W.H. Zurek, *Quantum Theory and Measurement* (Princeton University Press, Princeton, 1983)
16. S. Perlmutter et al., Measurements of Ω and Λ from 42 High-Redshift Supernovae. The Astrophys. J. **517**, 565 (1999)
17. L. Smolin, Did the universe evolve? Class. Quant. Gravity **9**, 173–191 (1992)
18. L. Smolin, *The Life of the Cosmos* (Oxford University Press, Oxford, 1997)
19. L. Smolin, The status of cosmological natural selection. arXiv:hep-th/0612185 (2006)
20. O.C. Stoica, Schwarzschild singularity is semi-regularizable. Eur. Phys. J. Plus **127**(83), 1–8 (2012).arXiv:1111.4837 [gr-qc]
21. O.C. Stoica, Big Bang singularity in the Friedmann-Lemaître-Robertson-Walker spacetime, December 2011. arXiv:1112.4508 [gr-qc]
22. O.C. Stoica, Beyond the Friedmann-Lemaître-Robertson-Walker Big Bang singularity. Commun. Theor. Phys. **58**(4):613–616, March 2012. arXiv:1203.1819 [gr-qc]
23. M. Tegmark, The mathematical universe. Found. Phys. **38**(2), 101–150 (2008)
24. M. Tegmark, *Our Mathematical Universe: My Quest for the Ultimate Nature of Reality*. (Knopf Doubleday Publishing Group, 2014)
25. Scott Aaronson. The ghost in the quantum turing machine. in *To appear in "The Once and Future Turing: Computing the World"*, a collection ed. by S. Barry Cooper, A. Hodges (2013) arXiv:1306.0159
26. S.W. Hawking, *A Brief History of Time* (Bantam Books, New York, 1998)

27. J. Barbour. Bit from it. Foundational Questions Institute, essay contest on "Is Reality Digital or Analog?" (2011) http://fqxi.org/community/essay/winners/2011.1barbour
28. O.C. Stoica, Toward a principle of quantumness. Preprint February 2014 arXiv:1402.2252 [quant-ph]
29. O.C. Stoica, Can the clicks of the detectors provide a complete description of nature? Preprint February 2014 arXiv:1402.3285 [quant-ph]
30. O.C. Stoica, Smooth quantum mechanics. PhilSci Archive (2008). http://philsci-archive.pitt.edu/archive/00004344/
31. O.C. Stoica, Flowing with a Frozen River. Foundational Questions Institute, "The Nature of Time" essay contest (2008) http://fqxi.org/community/forum/topic/322
32. O.C. Stoica, Modern physics, determinism, and free-will. Noema, Romanian Committee for the History and Philosophy of Science and Technologies of the Romanian Academy, XI: 431–456 (2012). http://www.noema.crifst.ro/doc/2012_5_01.pdf
33. O.C. Stoica, Convergence and free-will. PhilSci Archive (2008). philsci-archive:00004344/
34. G.D. Birkhoff, Universal algebra, in *Comptes Rendus du Premier Congrès Canadien de Mathématiques*, vol. 67 (University of Toronto Press, Toronto, 1946), pp. 310–326
35. C.C. Chang, H.J. Keisler, *Model Theory*, vol. 73. (North Holland, 1990)
36. O.C. Stoica, Matematica şi lumea fizică Noema, Romanian Committee for the History and Philosophy of Science and Technologies of the Romanian Academy, XIII:401–405 (2013). http://www.noema.crifst.ro/doc/2013_2_4.pdf

Chapter 6
Information-Based Physics and the Influence Network

Kevin H. Knuth

Abstract I know about the universe because it influences me. Light excites the photoreceptors in my eyes, surfaces apply pressure to my touch receptors and my eardrums are buffeted by relentless waves of air molecules. My entire sensorium is excited by all that surrounds me. These experiences are all I have ever known, and for this reason, they comprise my reality. This essay considers a simple model of observers that are influenced by the world around them. Consistent quantification of information about such influences results in a great deal of familiar physics. The end result is a new perspective on relativistic quantum mechanics, which includes both a way of conceiving of spacetime as well as particle "properties" that may be amenable to a unification of quantum mechanics and gravity. Rather than thinking about the universe as a computer, perhaps it is more accurate to think about it as a network of influences where the laws of physics derive from both consistent descriptions and optimal information-based inferences made by embedded observers.

An Electron Is an Electron Because of What It Does

As participants of the Information Age, we are all somewhat familiar with the electron. Currents of electrons flow through the wires of our devices bringing them power, transferring information and radiating signals through space. They tie us together enabling us to communicate with one another via the internet, as well as with distant robotic explorers on other worlds. Many of us feel like we have sensed electrons directly through the snap of an electric shock on a dry winter day or the flash and crash of a lightning bolt in a stormy summer sky. Electrons are bright, crackly sorts of things that jump and move unexpectedly from object to object. Yet they behave very predictably when confined to the wires of our electronic devices. But what are they really?

K.H. Knuth (✉)
Departments of Physics and Informatics, University at Albany,
Albany, NY 12222, USA
e-mail: kknuth@albany.edu

A. Aguirre et al. (eds.), *It From Bit or Bit From It?*,
The Frontiers Collection, DOI 10.1007/978-3-319-12946-4_6

Imagine that electrons could be pink and fuzzy. However, if each of these properties did not affect how an electron influences us or our measurement devices, then we would have no way of knowing about their pinkness or fuzziness. That is, if the fact that an electron was pink did not affect how it influenced others, then we would never be able to determine that electrons were pink. *Knowledge about any property that does not affect how an electron exerts influence is inaccessible to us.*

We can turn this thought on its side. *The only properties of an electron that we can ever know about are the ones that affect how an electron exerts influence.* Another way to think about this is that an electron does not do what it does because it is an electron; rather an electron is an electron because of what it does.

The conclusion is that *the only properties of an electron that we can know about must be sufficiently describable in terms of how an electron influences others.* That is, rather than imagining electrons to have properties such as position, speed, mass, energy, and so on, we are led to wonder if it might be possible, and perhaps better, to describe these attributes in terms of the way in which an electron influences. Since we cannot know what an electron *is*, perhaps it is best to simply focus on what an electron *does*.

The Process of Influence

Since we are aware of the existence of electrons, at the most fundamental level we can be assured that electrons exert influence. But we may wonder what such influence is like and whether there may be different types of influences. Most importantly, what exactly would we need to know about the process of influence to understand the electron?

Certainly it is conceivable that an electron could exert influence in a variety of ways. With this in mind, imagine that we have two electrons: one which influences in one way and another which influences in a different way. Since we identify and distinguish an electron from other types of particles (for lack of a better word) based on how it influences, we really have no way of telling if these are both electrons each exhibiting a different behavior from its repertoire, or whether these are simply two different types of particles altogether. Since we cannot possibly differentiate between the situation of two differently-behaving electrons and the situation of two different types of particles, such differentiation cannot affect any inferences we could make about the situation. Therefore we lose nothing by defining what we mean by an electron as being a particle that has only one particular way of influencing others. Now there are certainly other possibilities, but for the moment let us start with this simple idea and see what physics arises—adding complexity only when warranted. Here we make the basic **postulates** on which our influence model is based.[1]

[1] You may not like these assumptions—feel free to try others! For now, let's see what physics these give rise to.

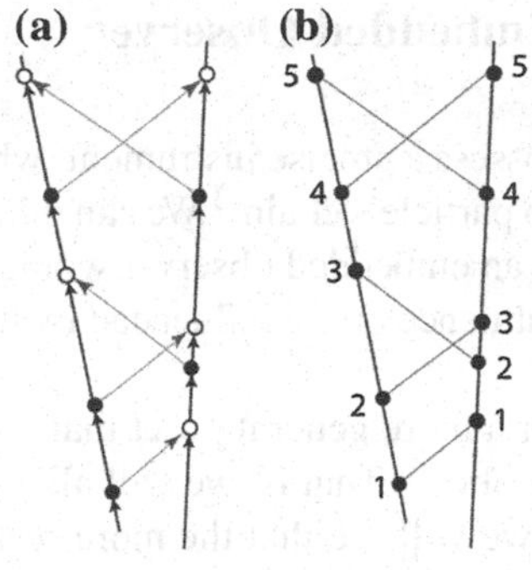

Fig. 6.1 **a** Illustrates an influence diagram of two interacting particle chains (*thick black lines*) that connect an ordered sequence of events. Influence is indicated by an *orange arrow* relating events representing acts of influence to responses to such influence. **b** Illustrates a Hasse diagram [1] of two particle chains. The *arrows* have been dropped with the understanding that lower events influence higher events along a connected path. The result is a partially-ordered set (poset) of events ordered by influence

#1. Particles can influence one another
#2. Influence is transitive (if A influences B and B influences C, then A influences C)
#3. Each instance of influence defines two events: the act of influence and response to influence
#4. For every pair of events experienced by a particle, one of these events influences the other

The result is that we have a set of events (#3), which potentially can be ordered by the process of influence (#2, #3, #4). Particles are described by an ordered sequence or *chain* of events (#4), which are mutually connected (#1) forming an acyclic graph, or a partially-ordered set (*poset* for short), which is analogous to what is called a causal set [2] or network where the events are *causally* ordered. We **do not** assume that these events take place in any kind of space or time.

Figure 6.1a illustrates two interacting particle chains with an *influence diagram* where the particle chains are indicated by the thick black lines that connect an ordered sequence of events, and influence is indicated by an orange arrow connecting one event representing an act of influence (black circle) on one chain to one other event representing the response to such influence (white circle) on a second chain. Each chain is conceptually analogous to a world line in relativity, though here a chain is a finite discrete structure, which does not reside in a pre-existing spacetime. For this reason, the directions of the chains, the fact that they are straight, and the distance between them on the page are not meaningful—only their connections matter. This diagram can be simplified into what is called a *Hasse diagram* (Fig. 6.1b) [1] by dropping the arrows and using height to indicate the direction of influence so that influence goes from the lower event to the higher event. We keep the thick lines to highlight the particle chains, and label the events with integers, whose order is isomorphic to the totally-ordered particle events.

Quantification by an Embedded Observer

We imagine an observer to possess a precise instrument, which has access to and can count the events along a given particle's chain.[2] We can think of this as the observer's clock. We may ask how such an embedded observer would describe this universe of events. Since not all events influence or are influenced by the observer, only a subset of events will be accessible.

To begin, we first consider a more general poset that allows greater connectivity than defined in our postulates above. That is, we will allow each event to connect to possibly many others. Later we will see that the more restricted connectivity gives rise to some quantum peculiarities; whereas the more general connectivity is more amenable to spacetime physics. The idea here is that we will develop a consistent observer-based scheme to quantify the poset of events based only on the numbers labeling the sequence of events along the embedded observer chain. We have shown that this quantification scheme is unique up to scale [3].

First, consider an observer chain P. Since the events that define the chain P are totally ordered and isomorphic to the set of integers under the usual ordering ($<$), we lose no generality by simply labeling (numbering) events with integers 1, 2, 3, etc. as was illustrated in Fig. 6.1b. Next we note that there exists a subset of events in the poset that *influence* events on the quantifying chain P. We say that such events *forward project* to the chain P. Similarly, there exists a subset of events that *are influenced by* events on the chain P. We say that these events *backward project* onto the chain P. This allows us to define a forward projection operator, P, that takes an event x that influences some elements on the chain and maps it to the *least* event on P that it influences, which we denote as Px. Similarly, we can define a backward projection operator, $\overline{P}$, that takes an event x that is influenced by some elements on the chain and maps it to the *greatest* event on P that influences it, which we denote as $\overline{P}x$. We can then label events in the poset based on the labels of the events that project to the chain P (Fig. 6.2). For example, an event x that both forward and backward projects to the chain P is quantified by the pair $(Px, \overline{P}x)$. The result is a chain-based coordinate system that covers part of the network.

We can now build up some extra structure by thinking about relations between events. Two events along a chain define an *interval*. For example, the interval denoted $[3, 5]$ along a chain is defined by the set of events $\{3, 4, 5\}$. Since combining intervals (set union) that share a common endpoint is associative, one can show that any non-trivial scalar measure of the interval must be additive [3]. This allows us to write the length of an interval as

$$d([x, y]) = y - x. \tag{6.1}$$

---

[2] We are not going to worry whether an event on the observer chain constitutes a measurement or detection.

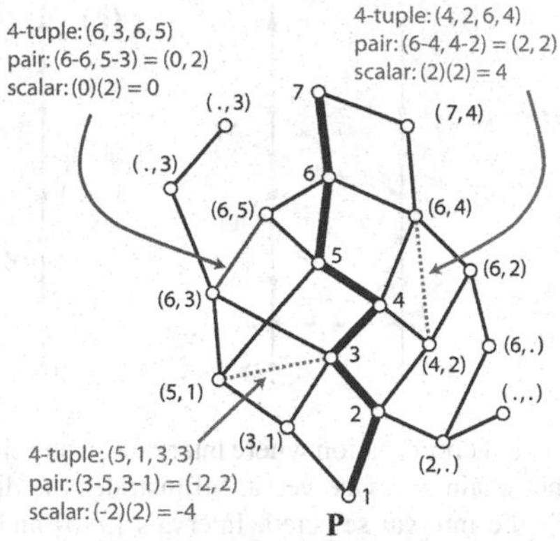

Fig. 6.2 A poset of events quantified by a chain P. Each event is quantified by at most two numbers. The first is found by forward projecting the event onto the quantifying chain P by identifying the least element on the chain P that is influenced by the event and the second is found by back projecting the event onto the chain P by identifying the greatest element on the chain that influences the event. Events on the chain project onto themselves so that event 3 is labeled by a *symmetric pair* (3,3) (not shown). Not all events can be quantified, nor are quantifications necessarily unique. The quantification of three intervals (*dotted* or *solid colored lines*) is also illustrated

Quantification with Pairs of Chains

Since a chain can at most assign a pair of coordinates to each event, the quantified set is essentially two-dimensional, while the poset itself is non-dimensional as it does not exist in a spacetime. To come up with a consistent quantification scheme, we will imagine two observers represented by finite chains P and Q that are *coordinated* in such a way that they agree on the quantification of each other's intervals. That is, an interval of length Δp on chain P forward projects to an interval of length Δq on chain Q as well as backward projects to an interval of length$\Delta \overline{q}$ on chain Q such that $\Delta p = \Delta q = \Delta \overline{q}$ (Fig. 6.3a). An interval of length Δp on chain P can be written in terms of the forward projections onto the two chains (since $\Delta p = \Delta q$) as

$$d\left([p_i, p_j]\right) = \frac{\Delta p + \Delta q}{2} \tag{6.2}$$

where $\Delta p = p_j - p_i$ and $\Delta q = Qp_j - Qp_i = \Delta p$. We can also consider a measure that quantifies the relationship between the two coordinated chains P and Q, which we will call the *distance*. Associativity with respect to considering relationships among multiple chains requires that this measure be additive [3]. In addition it must depend on the projection lengths Δp and Δq of an interval $[p_i, q_j]$ where p_i and q_j are arbitrary events on P and Q, respectively. Choosing the scale to agree with (6.2) gives

$$D\,(P, Q) = D([p_i, q_j]) = \frac{\Delta p - \Delta q}{2} \tag{6.3}$$

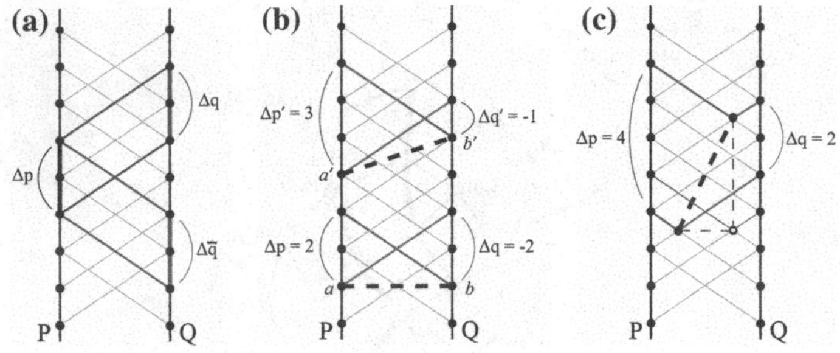

Fig. 6.3 **a** Illustrates the concept of coordination where intervals on one chain project onto intervals of the same length on the other chain and vice versa. **b** Illustrates the distance measure between chains. It does not depend on the interval selected. Intervals $[a, b]$ and $[a', b']$ are shown with distances $D(P, Q)$given by $(\Delta p - \Delta q)/2 = (2 - (-2))/2 = 2$ and $(\Delta p' - \Delta q')/2 = (3 - (-1))/2 = 2$. Note also that the interval $[a, b]$ is quantified by the *antisymmetric pair* $(2, -2)$ and the scalar $(2)(-2) = -4$, which is the reason for the minus sign in the metric (Eq. 6.5). **c** Illustrates the symmetric-antisymmetric decomposition. An interval quantified by the pair $(4, 2)$ is decomposed with an imaginary event (*open circle*) into an interval quantified by the pair $(3, 3)$ of length 3 along the chains and an interval quantified by the pair $(1, -1)$ with a distance of 1 between the chains so that $\Delta p\Delta q = (4)(2) = (3)(3) + (1)(-1) = 8$

where $\Delta p = Pq_j - p_i$ and $\Delta q = q_j - Qp_i$ for *any* event p_i on P and *any* event q_j on Q. This is illustrated in Fig. 6.3b.

We can generalize the concept of interval by considering a *generalized interval* $[a, b]$ defined by any two events a and b in the partially-ordered set. In the case where both a and b forward project onto chains P and Q, and are situated *between* P and Q (which is defined algebraically, see Appendix), we can quantify the interval in three ways [3] (other cases are similar):

$$\begin{array}{llll} (p_a, q_a, p_b, q_b) & & & \textit{quadruple} \\ (p_b - p_a, q_b - q_a) & \equiv & (\Delta p, \Delta q) & \textit{pair} \\ (p_b - p_a)(q_b - q_a) & \equiv & \Delta p \Delta q & \textit{scalar} \end{array}$$

where the scalar measure corresponds to a length squared (see Appendix). Any interval can be decomposed so that its pair is a component-wise sum of a *symmetric pair* of lengths along the chains (Fig. 6.2) and an *antisymmetric pair* of a distance between chains (Fig. 6.3b) in what we call the *symmetric-antisymmetric decomposition* (Fig. 6.3c) [3]

$$(\Delta p, \Delta q) = \left(\frac{\Delta p + \Delta q}{2}, \frac{\Delta p + \Delta q}{2}\right) + \left(\frac{\Delta p - \Delta q}{2}, \frac{\Delta q - \Delta p}{2}\right). \quad (6.4)$$

The scalar measure applied to each pair in this decomposition is also additive

$$\Delta p \Delta q = \left(\frac{\Delta p + \Delta q}{2}\right)^2 - \left(\frac{\Delta p - \Delta q}{2}\right)^2, \quad (6.5)$$

which is analogous to the Minkowski metric, $\Delta s^2 = \Delta t^2 - \Delta x^2$, with a 'time' coordinate $\Delta t = (\Delta p + \Delta q)/2$, which is defined by the ordering relation along chains, and 'space' coordinate $\Delta x = (\Delta p - \Delta q)/2$ defined by the induced ordering between chains [3]. Here "flat space" arises from a concept of influence in the case where we assumed that we could have coordinated chains that agree on the lengths of each other's intervals.[3] Since one ordering is natural (lengths), and the other induced (distance), we say that this is a 1 + 1-dimensional subspace. Note also that the proper time squared, Δs^2, is not actually a squared quantity in this picture since $\Delta s^2 = \Delta p \Delta q$.

We don't have to assume a condition as strong as coordination. We could instead assume that we have one chain that projects consistently to another, such that every interval of length $\Delta p = k$ on chain P forward projects to an interval of length $\Delta p' = m$ on P' and backward projects to an interval of length $\Delta q' = n$ on Q' (see Appendix). That is, an interval quantified by observers PQ as $(k, k)_{PQ}$ is quantified by observers P'Q' as $(m, n)_{P'Q'}$. We have shown that preserving the scalar measure leads to $k = \sqrt{mn}$ with the pair transformation [3]

$$(\Delta p', \Delta q')_{P'Q'} = \left(\Delta p\sqrt{\frac{m}{n}}, \Delta q\sqrt{\frac{n}{m}}\right)_{P'Q'} \tag{6.6}$$

which is related to the Bondi k-calculus [4] formulation of specialinteger quantifications. Changing variables to Δt and Δx and defining $\beta = \frac{\Delta p' - \Delta q'}{\Delta p' + \Delta q'} = \frac{m-n}{m+n}$ and $\gamma = \left(1 - \beta^2\right)^{-1/2}$ we obtain a Lorentz transformation analogue

$$\Delta t' = \gamma \Delta t - \beta\gamma \Delta x \quad \text{and} \quad \Delta \mathrm{x}' = -\beta\gamma \Delta \mathrm{t} + \gamma \Delta \mathrm{x}, \tag{6.7}$$

where the parameter β is analogous to *speed*.

At this point we have the ***poset picture*** where there only a network of influences—no physical spacetime and nothing moves. The projections, Δp and Δq, and ratio β describe how events and intervals relate to the observer chains. From this, we obtain an emergent ***spacetime picture*** where Δt and Δx assign times and positions to events, and the quantity β describes how the positions of successive events along a chain change.

We find also that β has a maximum invariant magnitude of one (analogous to the speed of light), which occurs whenever the projection m or n is zero. If we consider the intervals defined by an act of influence from one chain to another, we see that these correspond to $\beta = \pm 1$, so that in the spacetime picture influence "propagates" at a maximum speed. In the poset picture, this reflects the fact that information about influence traverses the network via transitivity, rather than defining a single event

---

[3] The signature of the metric, which determines where the minus sign goes, is in agreement with the particle physics tradition and opposite to that used in general relativity where one writes $\Delta s^2 = -\Delta t^2 + \Delta x^2$. Here the signature is not arbitrary since the minus sign comes from the fact that the interval between chains is quantified by a pair that has opposite signs. Later, this gives rise to the mass-energy-momentum relation with the correct signature.

for everything. To paraphrase Susan Sontag [5], another way to think about this is: "Time reflects the fact that everything does not happen at once, and space reflects the fact that not everything happens to you".

The Free Particle

Now let us go back to our particle model and consider what two coordinated observers would infer about a particle that is influencing them. In some sense, this is hokey because the observers are assumed to be coordinated, which means that they project to each other following connectivity rules that differ from those that our particle must follow where each event can connect at most two chains. This basically says that at a microscopic scale, we can never really have coordinated observers. Influence from any other particle will throw off our coordination. This is interesting, since that means that any external influence will ruin our nice flat Minkowski metric. This suggests that the influences that gave us our emergent flat spacetime also have the ability to curve it—potentially providing a route to quantum gravity. We may be able to achieve some kind of average coordination at larger scales by ignoring the tiny microscopic hiccups. Let's assume that this is the case and see what the observers would experience.

We define a *free particle* as a particle chain that influences others (according to our postulates), but is not itself influenced. We consider two observers that are influenced by this free particle, and record events generated by such influence as: p1, p3, p4, p6 and q2, q5, and q7 (Fig. 6.4a). While these detected events can be ordered on their respective chains, there is not enough information for the two observers to collectively reconstruct how the corresponding events were ordered along the particle

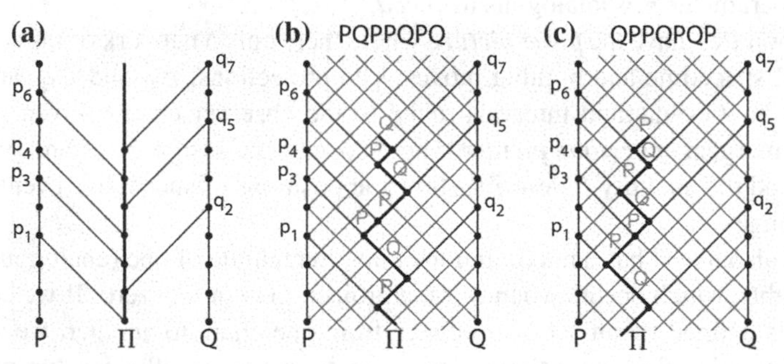

Fig. 6.4 **a** Illustrates the free particle Π in the poset picture as it influences coordinated observers P and Q. Each interval on Π projects to an interval of zero length on either P or Q, resulting in $\beta = \pm 1$. Furthermore, the observers have no way of determining the relative order of the P and Q events (for example, whether Π influenced P at p_1 first or Q at q_2 first). **b** Illustrates the correct reconstruction (PQPPQPQ) of the particle's influence pattern in the spacetime picture where time runs upward and the *horizontal* position in the picture indicates the position of the particle. The particle Π is observed to zig-zag at the speed of light. **c** Illustrates another possible reconstruction (QPPQPQP). Each of the 35 influence patterns corresponds to a discrete path in the emergent spacetime. Observer inferences must consider all possible reconstructions (spacetime paths)

chain. That is, despite the fact that the observers recorded all the information that is possibly available to them, it is impossible for them to definitively determine what the particle did. This *missing information* is an essential component of quantum mechanics, and here we see it arise from our simple model of influence. Let us look at this more closely and determine whether this can provide any meaningful insights into the quantum world.

BITs, ITs, and Fermion Physics

When we consider the set of influences {p1, p3, p4, p6} and {q2, q5, q7}, there are four interactions with the chain P and three interactions with the chain Q leading to $7!/(3!4!) = 35$ possible orderings along the particle chain Π, which we can list as PPPPQQQ, PPPQPQQ, ..., PQPPQPQ, ... QQQPPPP. These sequences represent all possible bit strings ($P \equiv 0$, $Q \equiv 1$) describing the particle's influences to the left and the right in this $1+1$-dimensional space. These sequences are constructed from the detection events (**Bit from It**) from which the observers must then make inferences (**It from Bit**) about the particle's behavior. In this sense, information is fundamental to the resulting physics. In the poset picture, these sequences correspond to all possible orderings of events along the particle chain. In the spacetime picture these correspond to all possible discrete spacetime paths, which are analogous to bishop moves on a chessboard.

Figure 6.4 shows two such reconstructions. It is instructive to consider how the intervals along the particle chain project directly onto one of the two observer chains so that it always has a projection of either $\Delta p = 0$ or $\Delta q = 0$, which means that $\beta = \pm 1$. That is, the particle is observed to zig-zag back-and-forth at the invariant speed (speed of light). This is an obscure quantum effect first proposed by Schrodinger in 1930, and only recently observed in the laboratory [10, 11], known as *Zitterbewegung* [6–8], which arises from the fact that the speed eigenvalues of the Dirac equation are $\pm c$ (the speed of light) [9].

We can consider inferences made by the observers about the particle's behavior. To compute probabilities [12], we must assign quantum amplitudes to each of the possible sequences and sum over them [13–15]. We can accomplish this with propagators that take the particle from some given initial state to a proposed final state. Figure 6.5 shows that given an assumed initial state in spacetime (x, t), there are two possible ways to have "arrived" there: from the left (P) and from the right (Q). These must both be considered. This also means that there are only two ways for a particle to *exist* at a given position at a given time, which is related to the familiar *Pauli Exclusion Principle*. While in three dimensions this involves the particle's *spin*, here in $1+1$-dimensions this involves its *helicity*, which simply is the direction of the previous influence event in the sequence. This suggests that spin is related to Zitterbewegung.

To make inferences using propagators, we need to keep track of four fundamental subsequences: PP, QP, QP, and QQ, whose probabilities sum to unity. These are

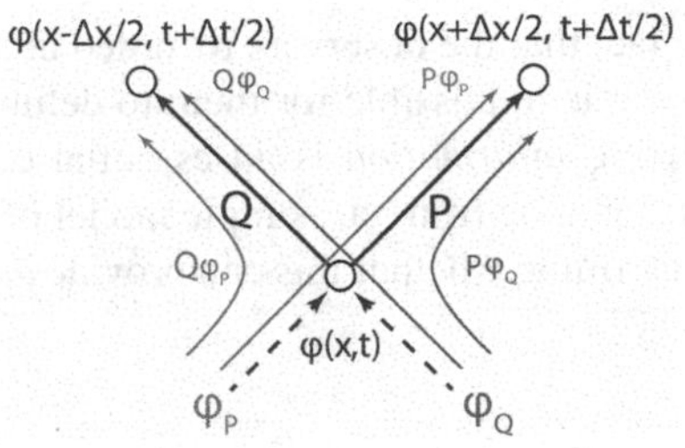

Fig. 6.5 Illustrates the two ways that a particle can arrive in an initial state (x,t) due to it having previously influenced P or Q. Complex numbers φ_P and φ_Q are assigned to each of the two initial sequence states and together comprise a Pauli spinor $\varphi(x,t) = \begin{pmatrix} \varphi_P \\ \varphi_Q \end{pmatrix}$. We have shown that the transfer matrices representing the propagator are given by $P = \frac{1}{\sqrt{2}}\begin{pmatrix} 1 & i \\ 0 & 0 \end{pmatrix}$ and $Q = \frac{1}{\sqrt{2}}\begin{pmatrix} 0 & 0 \\ i & 1 \end{pmatrix}$, which considers all four subsequences and accounts for Feynman's factor of i during helicity reversals [16]

encoded using a pair of complex amplitudes φ_P and φ_Q assigned to the initial states P or Q, which together comprise a Pauli spinor and are then propagated using two matrix operators P and Q, which take each of the two possible initial states to the two possible final states. This is starting to look a lot like the Dirac equation, and indeed we have shown [13] that this model is analogous to the Feynman checkerboard problem [16] where the Fermion is modeled as a particle that makes bishop moves on a chessboard. Feynman showed that by assigning an amplitude of $i\varepsilon$ for every direction reversal (helicity change), it is possible to obtain the Dirac equation in $1 + 1$ dimensions. We have *derived* Feynman's amplitude assignment with this model by observing that the probability associated with the sum of amplitudes $P\,\varphi_P + Q\,\varphi_P + P\,\varphi_Q + Q\,\varphi_Q$ is unity [13].

Mass, Energy and Momentum

Since the events are discrete, the emergent spacetime is discrete with a minimal dimension determined by the influence rate. Since an act of influence can result in a minimum of either $\Delta p = 1$ or $\Delta q = 1$, this corresponds to $\Delta t = +1/2$ and $\Delta x = \pm 1/2$, so that time always advances at least 1/2 a unit and the particle can go left or right by at least a 1/2 step. This makes time an excellent parameter for indexing observations. For an electron, these units could correspond to the Compton wavelength (where $\Delta t \approx 8 \times 10^{-21}$ s and $|\Delta x| \approx 2.4 \times 10^{-12}$ m).

So far, we have been considering inferences about intervals. We can also consider inferences about rates of influence, which is related to an internal electron clock rate first hypothesized by de Broglie in his 1924 thesis [8]. Let us define the rate at which the particle influences the chain P as $r_P = \#/\Delta p$ where # represents a given number of influencing events that are detected over an interval of length Δp. The rate r_Q

can be defined similarly. The product of the rates $r_P r_Q$, which is invariant since it is proportional to $(\Delta p \Delta q)^{-1}$, can be written as

$$r_P r_Q = \left(\frac{r_P + r_Q}{2}\right)^2 - \left(\frac{r_P - r_Q}{2}\right)^2, \tag{6.8}$$

which is analogous to the familiar mass, energy, momentum relation (in units where $c = 1$)

$$m^2 = E^2 - p^2 \tag{6.9}$$

where *mass* is analogous to the geometric mean of the rates of influence to the right and the left $m = \sqrt{r_P r_Q}$, *energy* is analogous to the arithmetic mean of the rates of influence $E = (r_P + r_Q)/2$, and *momentum* is analogous to the half-difference $p = (r_Q - r_P)/2$, which is defined with a sign change so that it agrees with the fact that as the particle influences more to the left, it is interpreted as moving to the right and vice versa. These defined quantities transform properly under the pair transformation (Lorentz transformation under a boost) and agree with the definition of β, which is analogous to speed:

$$\beta = \frac{p}{E} = \frac{r_Q - r_P}{r_P + r_Q} = \frac{\frac{\#}{\Delta q} - \frac{\#}{\Delta p}}{\frac{\#}{\Delta q} + \frac{\#}{\Delta p}} = \frac{\frac{\Delta p}{\Delta p \Delta q} - \frac{\Delta q}{\Delta p \Delta q}}{\frac{\Delta p}{\Delta p \Delta q} + \frac{\Delta q}{\Delta p \Delta q}} = \frac{\Delta p - \Delta q}{\Delta p + \Delta q} = \frac{\Delta x}{\Delta t} \tag{6.10}$$

The mass is related to the clock rate, which determines both the smallest time increment and distance that can be defined. It in this sense that mass is responsible for emergent spacetime.

Conclusion

It appears to be possible to obtain a great deal of physics as well as a number of particle "properties" from a simple model of an entity that influences others. Surprisingly we do not need to know *how* a particle influences others—just that it does—to obtain these relevant physical variables with their expected relations. This model of influence results in an emergent spacetime, which provides particles with positions at times, but we see that this breaks down in important quantum mechanical ways at the microscopic scale.

We also obtain insights into how mass, energy and momentum are related to rates (frequencies). We see that momentum cannot be defined simultaneously with position, since momentum is defined in terms of an average rate defined by a set of discrete influences, whereas position is defined (albeit with its own inherent uncertainty) by a pair of influences (one to the left and one to the right). At the microscopic level we do not have momentum, we have Zitterbewegung where the particle zig-zags at the maximum speed. The conceptual difficulties with quantum complementarity are eliminated when we consider these quantities to be *descriptions of what a particle does* rather than *properties possessed by the particle*.

The relation between reality (IT) and information about reality (BIT) comes into play twice in this model. First, “Bit from It” results from the fact that the particle does something (IT), which results in the observers recording detections (BIT). Second, “It from Bit” results from the fact that the observers make inferences about a set of relevant variables (“IT”) based on their information about detections (BIT). These relevant variables and their relations constitute a model (which we call physics) of a not completely knowable underlying reality. Rather than thinking about the universe as a computer, perhaps it is more accurate to think about it as a network of influences where the laws of physics derive from both consistent descriptions and optimal information-based inferences made by embedded observers.

Appendix

The key idea behind employing coordinated chains is that they provide a means of delineating a specific $1 + 1$-dimensional subspace in the non-dimensional poset. Events are defined to lie within the subspace defined by the two coordinated chains if the projection of the event onto one chain can be found by first projecting the event onto the other chain and then back to the first. This leads to a set of algebraic relations (for example $Px = \overline{P}Qx$ and $Qx = Q\overline{P}x$) where we consider the projections of event x onto chains P and Q. This in turn leads to several different relationships between an event x and the pair of chains (for example, x can be on the P-side of PQ, the Q-side of PQ or between PQ). For example, we say that event x is **between** two coordinated chains P and Q if $Px = P\overline{Q}x$, $\overline{P}x = \overline{P}Qx$, $\overline{P}x = \overline{P}Qx$, and $\overline{Q}x = \overline{Q}Px$ [3].

The derivation of the **scalar measure** is based on a *consistency requirement* that any two chains that agree on the lengths of each others intervals (coordination) must agree on the lengths of every interval that both chains can quantify. We assume that the scalar measure is a non-trivial symmetric function of the pairwise measure. That is, $s = \sigma(\Delta p, \Delta q) = \sigma(\Delta q, \Delta p)$, where $\sigma(\cdot, \cdot)$is a function to be determined. We can change our units of measure, so that we have $\alpha s = \sigma(\alpha\Delta p, \alpha\Delta q)$. This is a special case of the *homogeneity equation* [17]

$$F(zx, zy) = z^k F(x, y) \tag{6.11}$$

where in our problem the parameter $k = 1$. The general symmetric solution is given by $F(x, y) = \sqrt{xy}\ h(x/y)$, where h is an arbitrary function symmetric with respect to interchange of x and y. We can show that the function h is unity, and that lengths of intervals are given by $\sqrt{\Delta p \Delta q}$, which leads to the interval scalar $\Delta s^2 = \Delta p \Delta q$ [3]. We can next consider chains that are consistently related where every interval of length $\Delta p = k$ on chain P forward projects to an interval of length

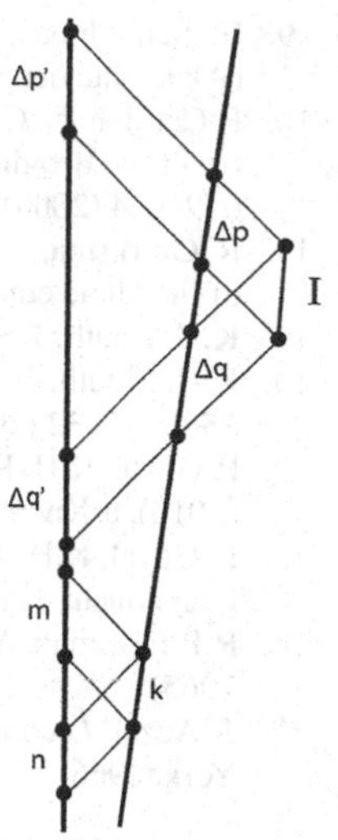

Fig. 6.6 Illustrates two consistently related chains. Chains Q and Q' are omitted. By coordination we have $\Delta\overline{p} = \Delta q$ and $\Delta\overline{p}' = \Delta q'$

$\Delta p' = m$ on P' and forward projects to an interval of length $\Delta q' = n$ on Q'. We now want to find a function L that takes the pair quantification of the interval I in the PQ frame to the P'Q' frame:$L_{PQ\rightarrow P'Q'}(\Delta p, \Delta q)_{PQ} = (\Delta p', \Delta q')_{P'Q'}$. (see Fig. 6.6). We note that we can write the projections of the interval I onto chain P in units of length k, so that the pairs can be written as $(\Delta p, \Delta q) = (\alpha k, \beta k)$and $(\Delta p', \Delta q') = (\alpha m, \beta n)$. Preserving the scalar measure gives $k^2 = mn$ so that $L_{PQ\rightarrow P'Q'}(\alpha k, \beta k)_{PQ} = (\alpha m, \beta n)_{P'Q'}$. It can then be shown that the general transform is given by $L_{PQ\rightarrow P'Q'}(x, y)_{PQ} = (x\sqrt{m/n}, y\sqrt{n/m})_{P'Q'}$ [3], which gives rise to the **Lorentz transformation** in (6.6) and (6.7).

References

1. B.A. Davey, H.A. Priestley, *Introduction to Lattices and Order* (Cambridge University Press, Cambridge, 2002)
2. L. Bombelli, J.-H. Lee, D. Meyer, R. Sorkin, Space-time as a causal set. Phys. Rev. Lett. **59**, 521–524 (1987)
3. K.H. Knuth, N. Bahreyni, A potential foundation for emergent space-time J. Math. Phys. **55**, 112501 (2014). http://dx.doi.org/10.1063/1.4899081, arXiv:1209.0881 [math-ph]
4. H. Bondi, *Relativity and Common Sense* (Dover, New York, 1980)
5. S. Sontag, *At the Same Time: Essays and Speeches* (Farrar Straus and Giroux, New York, 2007)
6. K. Huang, On the Zitterbewegung of the Dirac electron. Am. J. Phys. **20**, 479 (1952)
7. D. Hestenes, The Zitterbewegung interpretation of quantum mechanics. Found. Phys. **20**(10), 1213–1232 (1990)
8. D. Hestenes, Electron time, mass and Zitter. The Nature of Time Essay Contest, Foundational Questions Institute (2008)

9. E. Schrödinger, *Über die kräftefreie Bewegung in der relativistischen Quantenmechanik* (Akademie der wissenschaften in kommission bei W. de Gruyter u, Company, Berlin, 1930)
10. P. Catillon, N. Cue, M.J. Gaillard, R. Genre, M. Gouanère, R.G. Kirsch, M. Spighel, A search for the de Broglie particle internal clock by means of electron channeling. Found. Phys. **38**(7), 659–664 (2008)
11. R. Gerritsma, G. Kirchmair, F. Zähringer, E. Solano, R. Blatt, C.F. Roos, Quantum simulation of the Dirac equation. Nature **463**(7277), 68–71 (2010)
12. K.H. Knuth, J. Skilling, Foundations of inference. Axioms **1**, 38–73 (2012)
13. K.H. Knuth, Information-based physics: an observer-centric foundation. Contemp. Phys. **55**(1), 12–32 (2014). doi:10.1080/00107514.2013.853426. arXiv:1310.1667 [quant-ph]
14. P. Goyal, K.H. Knuth, J. Skilling, Why quantum theory is complex. Phys. Rev. A **81**, 022109 (2010). arXiv:0907.0909v3 [quant-ph]
15. P. Goyal, K.H. Knuth, Quantum theory and probability theory: their relationship and origin in symmetry. Symmetry **3**(2), 171–206 (2011)
16. R.P. Feynman, A.R. Hibbs, *Quantum Mechanics and Path Integrals* (McGraw-Hill, New York, 1965)
17. J. Aczel, *Lectures on Functional Equations and Their Applications* (Academic Press, New York, 1966)

Chapter 7
Relative Information at the Foundation of Physics

Carlo Rovelli

Second prize in the 2013 FQXi context "It From Bit or Bit From It?"

Abstract Shannon's notion of relative information between two physical systems can function as foundation for statistical mechanics and quantum mechanics, without referring to subjectivism or idealism. It can also represent a key missing element in the foundation of the naturalistic picture of the world, providing the conceptual tool for dealing with its apparent limitations. I comment on the relation between these ideas and Democritus.

Is There a Subjective Element in Statistical Mechanics?

Thermodynamical quantities such as entropy and temperature depend on the macroscopical variables chosen to describe systems with many degrees of freedom. They depend on coarse-graining. For instance, entropy can be defined (in the microcanonical) in terms of the number of microstates compatible with what we know about the system. With this definition, entropy changes if we know more. This appears to insert a puzzling subjective element in physics. There is a tension with the fact that termodynamical laws seem to hold quite independently on any choice or knowledge of ours. Is the Sun "hot" just because we "choose" a certain coarse graining for describing it? Does entropy increases because of our choices?

C. Rovelli (✉)
CPT, Université de Toulon, 13288 Marseille, France
e-mail: rovelli@cpt.univ-mrs.fr

C. Rovelli
CNRS UMR7332, Aix-Marseille Université, 13288 Marseille, France

A. Aguirre et al. (eds.), *It From Bit or Bit From It?*,
The Frontiers Collection, DOI 10.1007/978-3-319-12946-4_7

The way out of the puzzle is simple. Entropy is neither something inherent to the microstate of a system, nor something depending on our subjective "knowledge" about it. Rather, it is a property of certain (macroscopic) variables. For instance, the full state of a gas in a box is described by the position and velocity of its molecules. No entropy so far. But volume, total energy and (time averaged) pressure on the box boundaries are well defined functions of this state, and entropy is a function of *these*. This is the first step.

Now consider a situation where the gas interacts with a *second* system coupled only to volume, total energy and pressure of the gas (for instance, it interacts with the gas by a thermometer and a spring holding a piston). Then the physical interactions between the gas *and this system* are *objectively* described by thermodynamics.

In other words, it is not an arbitrary or subjective choice of a coarse-graining that makes thermodynamics physically relevant: it is the concrete way another physical system is coupled to the gas. If the coupling is such that it depends only on certain macroscopic variables of the gas, then the physical interactions between the gas and this second system are *objectively* well governed by thermodynamics.

This key observation clarifies the role that information plays in physics. Entropy, indeed, is information: in the micro-canonical language entropy is determined by the number of microstates compatible with a given macrostate. The number of states in which something can be, is precisely the definition of "information" (more precisely, "lack of information") given by Shannon in his celebrated 1948 work that started the development of information theory [1]. But "information", that is, the number of alternatives compatible with what we know, is not significative in physics insofar as it depends on idealistic subjective knowledge: it is relevant in physics when it refers to the *interaction between two systems* where the effects of the interaction on the second depend only on few variables of the first, and are independent on the rest of the variables. Under these circumstances, the number of states of the first system which are not distinguished by these variables is the number of Shannon "alternatives", relevant for the definition of thermodynamical entropy. Here "information", counts the number of states of a system which behave equally in the interaction with a second system.

Therefore the information relevant in physics is always the *relative* information between *two* systems. There is no subjective element in it: it is fully determined by the state *and* the interaction Hamiltonian which dictates which variables are the relevant ones in the interaction.

Pictorially: it is not the microstate of the Sun which is hot, it is the manner the Sun affects the Earth which is *objectively* hot.

Relative Irreversibility

Reconsider the quintessential irreversible phenomenon: a cup falls to the floor and breaks—in the light of the observation above. On one account this is obviously an irreversible phenomenon, but is it so on any possible account? The event is one among the many possible dynamical evolutions of a bunch of molecules. What makes the starting configuration more "special" that the final one? Something does so, but it is not in the microstate of the molecules: it is the manner we describe, or better, at the light of the previous section, we interact with it. It is because of our macroscopic account of the cup, dictated by the variables we interact with, that the initial state is special and therefore entropy increases.

To illustrate this, consider a box full of balls, characterized by two properties, say color and electrical charge. Say there are two possible colors: white and black; and two possible value of the charge: neutral and charged. Consider a microstate Col where white balls are on the left of the box and black balls on the right, while charge is randomly distributed. And consider a different microstate Ch, where charged balls are on the left and neutral balls on the right, while color is randomly distributed. To normal eyes, Col looks as a low-entropy state and Ch as a high-entropy state. But to a person who is color blind but has an electrometer it is Ch that looks low-entropy and Col that appears to have high entropy. Who is right? Both, of course. Entropy is relational: it pertain to the relation between two interacting systems, not to a single system.

Could the breaking cup be observed by somebody else, coupling differently to it, as a process where entropy decreases? Yes of course: imagine each fragment of the cup moving to a picture of itself on the ground, pictured in color to which you were color blind: you had missed the fact that the broken-cup state was very low entropy indeed…

If these considerations are correct, the irreversibility of the world is to be understood as a property of the couplings between systems, rather than a property of isolated systems. A conclusion that appears to run against what commonly said.

The Limits of Microphysics Without Information

The idea that the world can be described as a vast see of interacting atoms, and nothing else, can be traced to the ancient atomism of Democritus. The naturalistic and materialistic world view of Democritus was soon criticized by Plato and Aristotle on the ground that it fails to account for the forms, or the objects, that we see in the world. What makes a certain ensemble of atoms into a given object we recognize? Plato and Aristotle (in different manners) wanted to add "forms" to the naturalistic view of Democritus. For Plato, a horse is not just an aggregate of matter: it is an imprecise realization of the abstract form ("idea") of a horse. For Aristotle, the same horse is the union of its substance and its form. But if the form is something over and above the substance, what is it?

What is it that makes a random disposition of molecules into a cup? Which property of the Democritean atoms can generate collective variables? And how?

In fact, Democritus's idea was more subtle than everything being just atoms. Democritus says that three features are relevant about the atoms: the shape of each individual atom, the order in which they are disposed, and their orientation in the structure. And then he employees a powerful metaphor: like twenty letters of an alphabet can be combined in innumerable manners to give rise to comedies or tragedies, similarly the atoms can be combined in innumerable manners to give rise to the innumerable phenomena of the world.

But what is the relevance of the way in which atoms combine, in a world in which there is nothing else than atoms? If they are like letters of an alphabet, whom do they tell stories to?

I think that the key to the answer is in the observation in the first section: physical systems interact and affect one another. In the course of these interactions, the way one system happens to be leaves traces on the way another system is: correlations are established.

Following Shannon, we can say that a system S has information about a system S if there is a physical constraint such that the number of total states of the two systems is smaller than the product of the number of states of each. For instance: if the system s can be in the states a and b and the system S can be in the states A and B, but there is a physical constraint (say do to the way the two have interacted) that forbids the combinations (a, B) and (b, A), thus allowing only the two states (a, A) and (b, B), then we say that s has (one bit of) information about S. In words, if we see the state of s, we also know the state of S. Physical interactions determine constraints among systems: if a tree happen to fall on my head, then I cannot be standing smiling anymore: I have some information about the tree.

Thus, systems have information about one another, in the sense of Shannon. The lack of information that a system has about another is precisely the entropy of the second with respect to the first. It is relevant for the interactions with the first. It is the conventional thermodynamical entropy.

The other way around, the features of a system that *are* distinguished by the way they affect a second system define the *form* of the first system, in the sense of Plato and Aristotle. Forms are relative, and are determined by mutual interaction and mutual correlations. A horse is not a horse because of the microstate of its atoms: its being a horse is something that pertains the way it affects and is correlated to another physical system, for instance myself as a physical system.

Before pursuing this line of thinking, let me bring quantum theory into the picture.

Quantum Theory

The discovery of quantum theory has sharpened the role of information in our understanding of the world.

If we measure the state of a system with a certain precision, the resulting information specifies a region R of the phase space of the system. The unit of phase space volume is $length^2 \times mass \times time^{-1}$, namely $action$, per degree of freedom. In classical mechanics we can in principle refine measurement arbitrarily, therefore there is always a continuous (infinite) amount of missing information about a system, whatever the precision of the measurement.

No longer so after the discovery of quantum theory. For instance, if we measure the energy of a harmonic oscillator and we obtain the result that this is between E_1 and E_2, then there is only a *finite* number n of possible values that the energy can have. This is given by the area of the region of phase space included between the two surfaces E_1 and E_2, divided by the Planck constant (that is, $n = \frac{E_2 - E_1}{\hbar\omega}$).

This result is general: for all quantum systems, there is only a finite number of the orthogonal states per each finite region of phase space. The Planck constant determines the minimal phase space volume. Phase space volume measures the (missing) information we have about a system. It follows that quantum mechanics affirms that information is no longer continuous as in classical physics. It is discrete, and the Planck constant is the minimal unit of information.

This leads to a first principle at the basis of quantum theory:

I. Information in any finite region of the phase space of any system is finite.

This principle does not exhaust quantum theory, because it holds for any discrete classical system as well. What further characterises quantum theory is that information can become "irrelevant", and be renewed. If we have measured a system, the information we have about it allows us to predict its future. In quantum theory, we can always add *new* information to the state of a system, even after we have reached maximal information about it. By doing so, part of the old information becomes irrelevant. That is, is has no effect on future predictions. The typical case is a sequence of measurements of spins along different axes, in a two-state system. Each measurement brings novel information and makes the previous one irrelevant for the determination of the future.

This leads to the second principle at the basis of quantum theory:

II. It is always possible to acquire new information about a system.

The combination of these two principles generates the entire mathematical structure of quantum theory, up to some technical aspects, as was shown in [2]. Thus, *relative information* that systems have about one another is a key language for grounding quantum theory.

But what is information in this context? The answer is again in the observation of the first section. The meaning of "information" here is in the correlation that an interaction establishes between two systems. What is called a "measurement" in quantum mechanics, I believe, is simply a generic physical interaction that happens to establish a correlation between two systems. Then one system "knows" about the other, in the sense of Shannon's relative entropy. In quantum mechanical terms, a third system making a measurement on the second can immediately predict the outcome of a measurement she could perform on the first. Therefore the second system "has information" about the first. The full understanding of quantum theory and its apparent puzzles in these terms is called the "relational interpretation" of quantum mechanics. This was introduced in 1995 in [2] and opened the way to the current extensive use of information theory in the foundation of quantum theory.

In the relational interpretation, measurement is reduced to a normal physical interaction, but interactions between systems are exchanges of information, in the sense of Shannon's relative information, because they establish physical correlations between systems.

Quantum Gravity

Spacetime geometry is dynamical. If we do not disregard the dynamical aspects of spacetime geometry, then any physical system includes the spacetime region where it is located. This implies that there is natural identification between *physical systems* and *spacetime regions*.

If we, furthermore, do not disregard quantum theory, then we must take into account the fact that any interaction between spacetime regions is a quantum interactions between systems. Therefore there is exchange of informations across spacial regions.

But quantum interactions are quantized and discrete, because of the first postulate above. The quantum discreetness, combined to the fact that geometry is dynamical and therefore quantized, leads immediately to the discretization of space, idea that can be traced back to the thirties [3, 4] and has been concretized more recently. Discreetness is expressed by the discreetness of the area of two-dimensional surfaces [5, 6], which is the central result of loop quantum gravity.

The discreteness of the area is a reflection of the discreetness of the quantum information that can be transmitted across these surfaces. In particular, the finiteness of the entanglement entropy density across any surface is a consequence of the discretization of the area. This analog to the fact that the finiteness of the entropy of a black body cavity is a consequence of the discrete nature of the photons.

Quantum correlations established across surfaces are ubiquitous in quantum field theory. In fact, the existence of short scale quantum correlations between specially separated points is a property of the quantum field theory vacuum as well as all Fock states. It is tempting to reverse this relation and try to interpret spacetime topology as the manifestation of these quantum correlations. From this persecutive, the texture itself of spacetime could perhaps be understood as a result of the net of relative information established by quantum entanglement [7, 8]. I think that our ideas are only at the beginning in this direction.

Reality and Information

It seems to me that this ensemble of considerations conspire towards a picture where the fog begins a bit to dissipate over the intriguing role of information at the foundation of physics.

Information that physical systems have about one another, in the sense of Shannon, is ubiquitous in the universe. It has the consequence that on top of the microstate of a system we have also the informational state that a second system O has about any system S.

The universe is not just the position of all its Democritean atoms. It is also the net of information that all systems have about one another. Objects are not just aggregate of atoms. They are configurations of atoms singled out because of the manner a given other system interacts with them. An object is only such with respect to an observer interacting with it.

Among all systems, living ones are those that selection has led to persist and reproduce by, in particular, making use of the information they have about the exterior world. This is why we can understand them in terms of finality and intentionality. They are those that have persisted thanks to the finality in their structure. Thus, it is not finality that drives structure, but selected structures define finality. Since the interaction with the world is described by information, it is by dealing with information that these systems most effectively persist. This is why we have DNA code, immune systems, sensory organs, neural systems, memory, complex brains, language, books, MAC's and the ArXives. To maximize the management of information.

The statue that Aristotle wants made of more than atoms, is made by more than atoms: it is something that pertains to the interaction between the stone and brain of Aristotle, or ours. It is something that pertains to the stone, the goddess represented, Phidias, a woman he met, our education, and else. The atoms of that statue talk to us precisely in the same manner in which a white ball in my hand "says" that the ball in your hand is also white, if the two are correlated. By carrying information.

This is why, I think, from the basis of genetics, to the foundation of quantum mechanics and thermodynamics, all the way to sociology and quantum gravity, the notion of information has a pervasive and unifying role. The world is not just a blind wind of atoms, or general covariant quantum fields. It is also the infinite game of mirrors reflecting one another formed the correlations among the structures made by the elementary objects. To go back to Democritus metaphor: atoms are like an alphabet, but an immense alphabet so rich to be capable of reading itself and thinking itself. In Democritus words:

"The Universe is change, life is opinion that adapt itself".

References

1. C.E. Shannon, A mathematical theory of communication. Bell Syst. Tech. J. **27**(3), 379 (1948)
2. C. Rovelli, Relational quantum mechanics. Int. J. Theor. Phys. **35**(9), 1637 (1996). 9609002
3. M.P. Bronstein, Kvantovanie gravitatsionnykh voln (Quantization of Gravitational Waves). Zh. Eksp. Tear. Fiz. **6**, 195 (1936)
4. M.P. Bronstein, Quantentheorie schwacher gravitationsfelder. Phys. Z. Sowjetunion **9**, 140–157 (1936)
5. C. Rovelli, L. Smolin, Discreteness of area and volume in quantum gravity. Nucl. Phys. **B442**(593–622), 9411005 (1995)

6. A. Ashtekar, J. Lewandowski, Quantum theory of geometry. I: Area operators. Class. Quantum Gravity **14**, A55–A82 (1997), 9602046
7. E. Bianchi, R.C. Myers, On the architecture of spacetime geometry (2012) 1212.5183
8. E. Bianchi, H.M. Haggard, C. Rovelli, The Boundary is mixed (2013), 1306.5206

Chapter 8
Information and the Foundations of Quantum Theory

Angelo Bassi, Saikat Ghosh and Tejinder Singh

This essay received the second prize in the FQXi 2013 Essay Contest 'It from Bit, or Bit from It?'

Abstract We believe that the hypothesis 'it from bit' originates from the assumption that probabilities have a fundamental, irremovable status in quantum theory. We argue against this assumption and highlight four well-known reformulations/modifications of the theory in which probabilities and the measuring apparatus do not play a fundamental role. These are: Bohmian Mechanics, Dynamical Collapse Models, Trace Dynamics, and Quantum Theory without Classical Time. Here the 'it' is primary and thc 'bit' is derived from the 'it'.

Introduction

In the standard approach to quantum theory, the state of the quantum system is described by the wave function, whosc evolution is given by the deterministic Schrödinger equation. However, when a measurement is performed on this system by a classical apparatus, the outcome is not deterministically related to the initial state. Instead, in any specific realization of the measurement, one or the other outcome occurs with a certain probability.

A. Bassi(✉)
Department of Physics, University of Trieste, Strada Costiera 11, 34151 Trieste, Italy
e-mail: bassi@ts.infn.it

S. Ghosh
Department of Physics, Indian Institute of Technology Kanpur, Kanpur 208016, India
e-mail: saikatghosh@bose.res.in

T. Singh
Tata Institute of Fundamental Research, Homi Bhabha Road, Mumbai 400005, India
e-mail: tejinder.tifr@gmail.com

A. Aguirre et al. (eds.), *It From Bit or Bit From It?*,
The Frontiers Collection, DOI 10.1007/978-3-319-12946-4_8

This sudden onset of probabilities in a system which is otherwise evolving deterministically while it is 'unmeasured', has sometimes been used to accord probabilities a fundamental, **irremovable** status in quantum theory. It has been suggested that the system or the object being measured upon (say an electron) does not have definite properties before the measurement, but rather resides in some probabilistic realm, and acquires well-defined physical properties only upon measurement (an act of seeking information). This outlook, namely that **the definitive properties of a quantum system somehow become a reality only when information about them is sought by an act of measurement**, is perhaps the foundation of the hypothesis 'it from bit'.

If we dwell on the above reasoning, it does not take much effort to narrow down to two places where the argument is weak enough to be essentially flawed:

- One is the so-called classical measuring apparatus, and the other is the status of probabilities. When is an apparatus classical? Strictly speaking, we do not quite know. Quantum theory does not say how large an object must be, before it can be said to obey the rules of Newtonian mechanics. Should its mass be a billion a.m.u. or a trillion a.m.u. or something else? The theory does not tell us. And why should the theory have to depend on its own limit [the classical apparatus, whatever that might mean] in order to complete the description of the formalism? Indeed, when something so well-defined mathematically such as the Schrödinger equation and the commutation relations are supplemented by something as vague as a 'measuring apparatus', we should smell rat, and know that things are amiss! We should look for a first principles holistic description of quantum theory which does not make explicit reference to a *classical* measuring apparatus.
- And secondly, there is no place for probabilities in a system evolving deterministically, and for which the initial conditions [the wave function] are exactly known. Probabilities arise when there is a pre-given sample to choose from, and the initial state is not precisely known. Such of course is the case in statistical mechanics, and in coin tossing. But not so in quantum theory—the only known physical theory where probabilities come into play without there being a sample of initial states.

The only way to overcome this illogical state of affairs is to look for a more complete formulation of the theory which offers a mathematical explanation for the random outcomes of measurements, thus getting rid of the fundamental inexplicable status of probability.

This could be achieved if the evolution is deterministic but the initial state is not precisely known [**Bohmian mechanics**]. Or it could happen if there is a stochastic nonlinear aspect to the evolution, over and above the Schrödinger evolution, which becomes significant during measurement, and dynamically causes collapse of the wave function into one or the other outcomes, in accordance with the Born probability rule [**dynamical collapse models**]. Remarkably enough, Bohmian mechanics as well as collapse models also do away with any direct reference to the classical measuring apparatus. A measurement is nothing but another aspect of dynamical evolution.

Figure 8.1 makes it abundantly clear that one should not attribute the emergence of random outcomes from a deterministic evolution to some mysterious and mathematically ill-defined 'measurement'. The evolution should either be supplemented

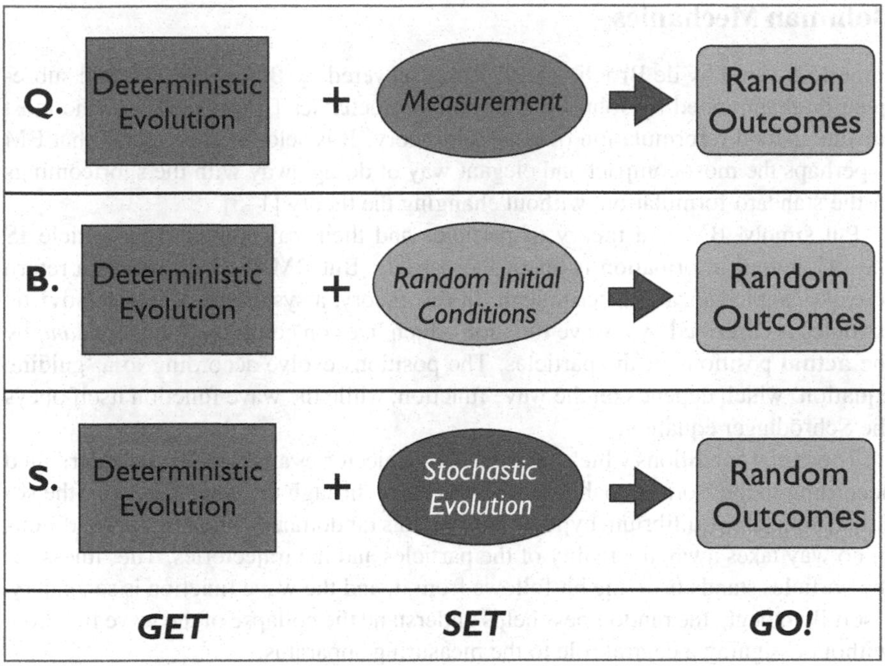

Fig. 8.1 The quantum *Q*. gets a makeover: *B*. Bohmian Mechanics *S*. Stochastic + Deterministic Evolution

by random initial conditions [**B.** Bohmian Mechanics] or by a stochastic aspect to the evolution itself [**S.** Schrödinger + Stochastic Evolution]. Such restoration of mathematical completeness is how physicists generally go about amending shortcomings in a physical theory.

When such a new formulation or a modified theory is presented, the 'it from bit' vanishes into thin air, literally! No longer does the existence or reality of physical systems depend on measurements/information/the questions we ask about them. The 'it' necessarily comes first. Below, we recall four different routes towards a more complete formulation of quantum theory, where neither the classical apparatus nor probabilities play a fundamental role, and in each of the routes, the 'it' regains its primary status.

It is sometimes said that it is not necessary to consider reformulations/modifications of quantum theory, because the theory agrees with every experiment performed to date, to test it. While this agreement of theory and experiment is undoubtedly remarkable, one should not forget that there are regions of the parameter space (quantified by number of degrees of freedom in the object being studied) where the theory has not yet been subjected to laboratory tests. Such ongoing/planned tests hold the potential for revealing the need to modify the theory.

Bohmian Mechanics

First discovered by de Broglie in 1927, rediscovered by Bohm in 1952, and subsequently championed by John Bell, Bohmian Mechanics [BM] is perhaps the most misunderstood reformulation of quantum theory! It is seldom appreciated that BM is perhaps the most compact and elegant way of doing away with the shortcomings of the standard formulation, without changing the theory [1–5].

Put simply, BM is a theory of particles and their trajectories. The particle **IS** the 'it' and all information is about the particle. But BM is by no means a return to classical physics and determinism. In this theory, a system of N non-relativistic particles is described by a wave function which lives on configuration space, *and* by the **actual** positions of the particles. The positions evolve according to a 'guiding equation' which depends on the wave function, while the wave function itself obeys the Schrödinger equation.

The initial conditions which determine the trajectories are random, and distributed according to the Born probability rule expressed through the wave function [the so-called quantum equilibrium hypothesis]. But this randomness of the initial conditions in no way takes away the reality of the particles and the trajectories. The 'itness' of the particles stands firm, the bit follows from it, and the wave function is secondary. Usefully though, the randomness helps understand the collapse of the wave function, without assigning a central role to the measuring apparatus.

BM explains everything that conventional non-relativistic QM does. But then people ask: what good is a reformulation which gives the same results as the standard theory? [A renowned physicist has gone as far as to call BM 'verbal window dressing'.] Such a question would perhaps not be asked if BM had been discovered before the standard theory replete with the Copenhagen interpretation. Instead we might be asking ourselves: what good is the Copenhagen interpretation and the accompanying ad hoc probabilities when we already have BM?!

It is undeniably the case that the chronological order in which different formulations were discovered, and their accompanying sociological impact, has had a great deal to do with the bit preceding the it, in some quarters.

Dynamical Collapse of the Wave Function

Or it could be that the fundamental evolution is nonlinear and stochastic, which includes the collapse of the wave function together with the standard quantum properties. How could we be so certain that in quantum theory, there is only a linear, deterministic, part to the evolution, described by the Schrödinger equation? The history of science has shown that linear theories often are approximations to more fundamental nonlinear theories, like in the case of Newtonian gravitation and general relativity. Therefore it seems natural to seek out nonlinear extensions of the Schrödinger equation.

There is nothing in today's experiments which rules out the inclusion of a stochastic nonlinear aspect in the quantum evolution. All that is required is that such an aspect should be extremely tiny and negligible for microscopic systems. On the other hand the stochastic aspect can well become significant for macroscopic systems. This is not ruled out by experiments; on the contrary, the random nature of outcomes in a measurement suggests such a feature!

This is because stochastic nonlinearity breaks quantum linear superposition during a measurement. When a quantum system interacts with a measuring apparatus, together they behave like a macroscopic system, for which the stochastic nonlinear evolution dominates the linear Schrödinger evolution, resulting in random outcomes which do not obey linear superposition, as observed.

In such a scenario, the 'it' is once again primary, being an objective reality described by the quantum state of the system. It is no problem that the evolution of this state is described by a stochastic Schrödinger equation. Once one has the collapse in the dynamics, the wave function provides a satisfactory description of physical reality.

The derived 'bit' is constituted by all the information we have about this quantum system—conventionally such information is collected through the interaction of this system with a macroscopic object obeying the laws of classical mechanics. It is no surprise if the information we collect is based on outcomes which are probabilistic, for this is an inevitable consequence of the interaction of a highly deterministic [quantum] system with a highly stochastic [classical] system.

Such a modified quantum theory, which combines deterministic evolution with stochastic evolution, has been successfully developed by a group of physicists, since the eighties [6–9]. In its currently most advanced version it is known as 'Continuous Spontaneous Localization' [CSL] [10–12]. This is a stochastic nonlinear Schrödinger equation, in which the standard linear evolution is supplemented by a nonlinear stochastic part, described by a Weiner process. The stochastic part enforces significant consequences on the Schrödinger evolution. It causes position localization by opposing the quantum spread of the wave function. Such localization is shown to be unimportant for micro-systems, thus explaining their wavy nature. On the other hand the localization is very significant for macro-systems, thus explaining their classical Newtonian behaviour. There is hence a universal dynamics, which on the one hand can explain the quantum nature of atomic and molecular phenomena, and on the other hand explain the classical nature of large objects. And none of this makes any reference to measurement.

Nor does this universal dynamics leave any place for fundamental, irremovable probabilities. The Born probability rule is shown to be a mathematical consequence of the CSL equation. When a quantum system, which is in a superposition of various eigenstates of the observable being measured, interacts with a classical measuring apparatus, the CSL equation shows that superposition is broken and one or the other outcomes is realized in accordance with this probability rule. This is random determinism: randomness is a fundamental feature of the law of evolution; it does not in any way take away the primary importance of the 'it'.

Today, technology is at a stage where CSL is being put to experimental tests in the laboratory, for in the mesoscopic and macroscopic range its predictions markedly differ from those of quantum theory [12]. If CSL is confirmed, then 'it' will reign supreme. If CSL is ruled out, Bohmian mechanics will gain centerstage, and 'it' will still hold ground!

Of course, having introduced a stochastic element into natural laws, the onus is on CSL to explain where this stochasticity comes from. Else, the criticism that it was invented for the sole purpose of explaining measurements and removing probabilities would be well-founded!

Perhaps there is a universal stochastic field in nature, of cosmological and/or gravitational origin. This field interacts with all material objects in the manner described by CSL. In fact the universal nature of fluctuations of the gravitational field, and their consequent impact on quantum evolution, has been emphatically highlighted. And this idea is also being put to test in the laboratory.

Alternatively—and this makes the case for 'bit from it' ever stronger—the stochastic element is a consequence of coarse graining of a fundamental deterministic theory which describes the evolution of the state of the 'it'. Such a theory, from which CSL originates, and to which quantum theory is an approximation, has indeed been developed by Stephen Adler and colaborators, and we briefly allude to it below.

Trace Dynamics

Why should there be any such thing as 'quantization'? Why do we have to be first given a classical theory, and we then use a recipe to obtain the quantum theory by 'quantizing' the classical theory? If we have to use a theory's own limit to deduce the theory, this is not the most satisfactory state of affairs, and this is one of the problems Adler's well thought out theory of Trace Dynamics [TD] sets out to address [13–16]. TD is a theory of the classical dynamics of Grassmannian matrices. A matrix degree of freedom at a given point in space may be thought of as representing a particle, and its time evolution, given by Newtonian dynamics, defines a spacetime 'trajectory' for the matrix. This matrix is the ultimate 'it', and if there were any misgivings caused by the stochasticity explicitly introduced by hand in CSL, those misgivings are removed here. For there is no fundamental stochasticity here, to begin with.

Why does one start with matrices? Because these matrices, which do not commute with each other, and have arbitrary commutation relations, serve as precursors of the position and momentum operators of quantum theory. One of the most remarkable and unique features of this matrix theory is that it possesses a conserved charge, made out of the sum of the commutators of the matrices and their corresponding momenta. Even though each of these commutators is arbitrary and time-dependent, their sum is conserved! This charge, which has the dimensions of action, plays a central role in the deduction of quantum theory from TD.

Next, one posits that at the scale at which we perform our laboratory experiments, we do not probe these matrices. Much in the same way in which while studying the macroscopic properties of a gas, we do not probe individual atoms, but only the thermodynamic properties of the coarse-grained system. In the same spirit one constructs the statistical thermodynamics of the dynamical theory of matrices: the matrices are the atoms, and their statistical averaging is the macroscopic gas.

Following the well laid-out laws of equilibrium statistical mechanics, one constructs a probability density distribution, whose equilibrium configuration is derived by maximizing the Shannon entropy. Given this probability distribution, and the invariance of thermodynamic averages under translations in phase space, one is led to an analog of the equipartition theorem. From here emerge the canonical commutation relations of standard quantum theory, satisfied by thermal averages of the underlying matrices (operators). There also emerge the Heisenberg equations of motion satisfied by the thermally averaged position and momentum operators. As in standard quantum theory, there is a Schrödinger evolution equivalent to the Heisenberg evolution. **Quantum theory is the statistical thermodynamics of a classical matrix dynamics.**

Next comes the icing on the cake: we find the 'bit' emerging from the 'it'. Where there is equilibrium thermodynamics, there are statistical fluctuations [Brownian motion]. While quantum theory corresponds to the equilibrium thermodynamics of the averaged matrix theory, the inclusion of fluctuations results in a modified nonlinear stochastic Schrödinger equation of the CSL type! Probabilities are thus the consequence of a stochastic element which has emerged from coarse graining an underlying deterministic theory. One could not be witness to a more convincing demolition, than this one, of the 'it from bit' hypothesis.

It is noteworthy that Bohmian Mechanics, Spontaneous Localization and Trace Dynamics are three experimentally distinguishable 'its' leading to the same 'bit' (Fig. 8.2). While BM makes the same experimental predictions as standard quantum

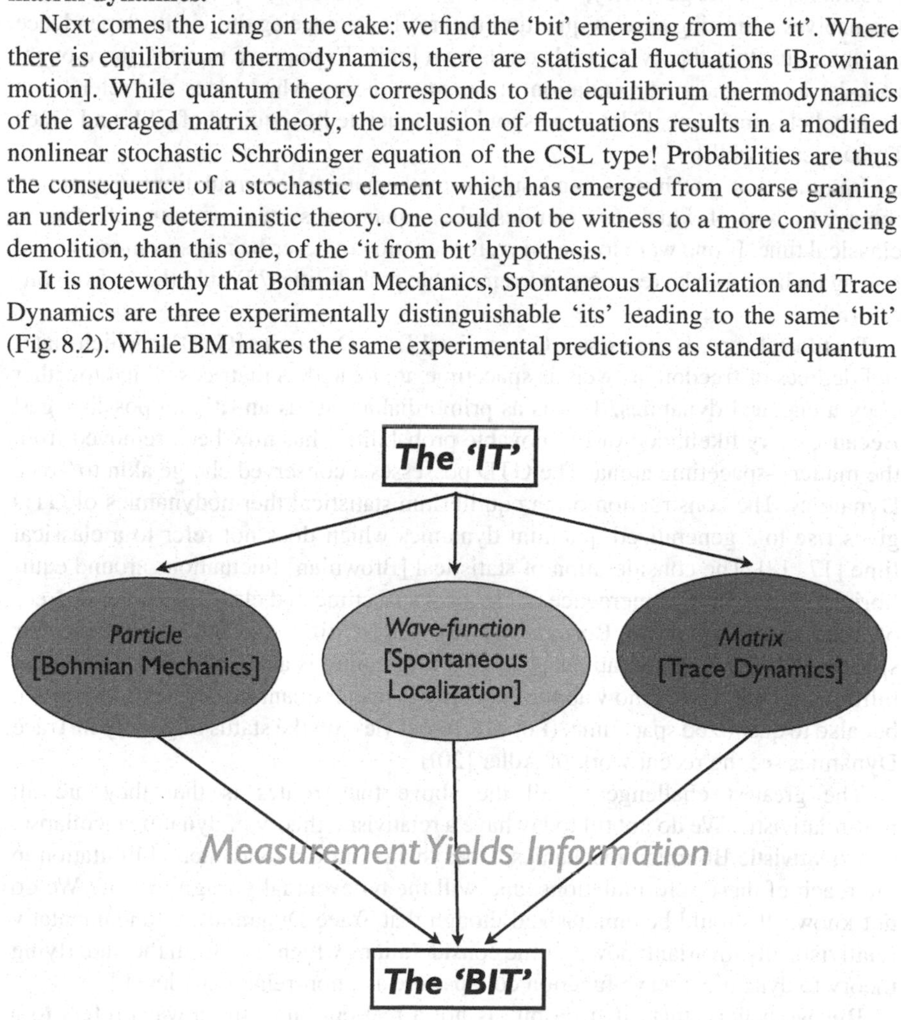

Fig. 8.2 Bit from It: The Threefold Way: Bohmian Mechanics/Spontaneous Localization/Trace Dynamics

theory, the predictions of CSL differ from those of quantum theory in the mesoscopic domain. TD agrees with CSL at low energies but differs from it at energy scales approaching the Planck domain.

Quantum Theory Without Classical Time

The story would not be complete until one last barrier has been crossed. Quantum theory as we know it, depends on an external classical time. Such a time belongs to a classical spacetime geometry, which is produced by macroscopic bodies, which are themselves a limiting case of quantum theory. Thus, once again, via its dependence on time, quantum theory depends on its own limit. To improve matters, there ought to exist an equivalent formulation of quantum theory which does not refer to an external classical time. [The same should in principle be required of CSL and Trace Dynamics as well.]

Imagine that one has at hand such a mathematical reformulation of quantum theory, by way of which one can describe a quantum system without referring to classical time. If one were to use this reformulation to describe measurement, would one again have to take recourse to irremovable probabilities? Would the 'it from bit' return?

Fortunately not. There exists a Generalized Trace Dynamics [GTD] in which material degrees of freedom as well as spacetime are treated as matrices which together obey a classical dynamics. This is as primordial an 'it' as an 'it' can possibly get! Because every likelihood of irremovable probabilities has now been removed from the matter—spacetime arena. The GTD possesses a conserved charge akin to Trace Dynamics. The construction of an equilibrium statistical thermodynamics of GTD gives rise to a generalized quantum dynamics which does not refer to a classical time [17, 18]. The consideration of statistical [Brownian] fluctuations around equilibrium allows for the emergence of classical spacetime and classical matter degrees of freedom, as well as the Born probability rule for different random outcomes for spacetime-matter configurations [19]. Coarse graining is again at the root of probabilities, and 'bit from it' now applies not only to matter quantum degrees of freedom, but also to quantized spacetime. (For a different view on the status of gravity in Trace Dynamics see the recent work of Adler [20]).

The greatest challenge to all the above four routes is that they are all non-relativistic. We do not till today have a relativistic theory of dynamical collapse, or a relativistic Bohmian mechanics. Does this point to a fundamental limitation in the reach of these reformulations, and will the bit eventually reign over it? We do not know. [It should be emphasized though that Trace Dynamics is fundamentally relativistically invariant; however the considerations which lead from the underlying theory to dynamical wave function collapse are at a non-relativistic level.]

But we believe that "it from bit" is not a real option. "Bit" always refers to a pre-existing "it" (Fig. 8.2). This is the meaning of "bit". All confusion comes from inverting the order of "bit" and "it". When one starts the right way with the "it", then all problems evaporate.

This work is supported by grants from the Foundational Questions Institute and the John Templeton Foundation.

References

1. D. Bohm, Phys. Rev. **85**, 166 (1952)
2. D. Bohm, Phys. Rev. **85**, 180 (1952)
3. J. Bub, *Interpreting the Quantum World* (Cambridge University Press, Cambridge, 1997)
4. D. Dürr, S. Teufel, *Bohmian Mechanics* (Springer, Heidelberg, 2009)
5. P.R. Holland, *The Quantum Theory of Motion: An Account of the de Broglie-Bohm Causal Interpre-tation of Quantum Mechanics* (Cambridge University Press, Cambridge, 1993)
6. P. Pearle, Phys. Rev. D **13**, 857 (1976)
7. L. Diósi, Phys. Rev. A **40**, 1165 (1989)
8. G.C. Ghirardi, A. Rimini, T. Weber, Phys. Rev. D **34**, 470 (1986)
9. N. Gisin, J. Phys. A **14**, 2259 (1981)
10. G.C. Ghirardi, P. Pearle, A. Rimini, Phys. Rev. A **42**, 78 (1990)
11. A. Bassi, G.C. Ghirardi, Phys. Rep. **379**, 257 (2003)
12. A. Bassi, K. Lochan, S. Satin, T.P. Singh, H. Ulbricht, Rev. Mod. Phys. **85**, 471 (2013)
13. S.L. Adler, Nucl. Phys. B **415**, 195 (1994)
14. S.L. Adler, *Quantum Theory as an Emergent Phenomenon* (Cambridge University Press, Cambridge, 2004), p. xii+225
15. S.L. Adler, J. Phys. A **39**, 1397 (2006)
16. S.L. Adler, A.C. Millard, Nucl. Phys. B **473**, 199 (1996)
17. K. Lochan, T.P. Singh, Phys. Lett. A **375**, 3747 (2011)
18. K. Lochan, S. Satin, T.P. Singh, Found. Phys. **42**, 1556 (2012)
19. T.P. Singh, in *The Forgotten Present (in press)*, ed. by T. Filk, A. von Muller (Springer, Berlin, 2013). arXiv:1210.8110
20. S. L. Adler (2013) arXiv:1306.0482

Chapter 9
An Insight into Information, Entanglement and Time

Paul L. Borrill

Abstract We combine elements of Boltzmann's statistical account of thermodynamic processes in the second law, Poynting's *twist waves* on a photon shaft and Shannon's theory of communication within a background-free conceptualization of *time*; where the departure and arrival of *information* carried by photons bounds "elements of physical reality" as perpetually reversible photon links embedded in an entangled network. Entangled networks become progressively irreversible as decoherence ebbs and flows with the environment. From this, we can begin to formulate a new and logically consistent view of the apparent non-locality revealed in violations of Bell's inequality.

Introduction

> Church's thesis and the Turing machine are rooted in the concept of *doing one thing at a time*. But we do not really know what doing is—or time—without a complete picture of quantum mechanics and the relationship between the still mysterious wave-function and macroscopic observation.
>
> –*Andrew Hodges in:*
> *Alan Turing: Life and Legacy of a Great Thinker* [1]

Our argument brings a new information-theoretic quality to the nature of an *interaction*. A perpetually alternating exchange of information between atoms by a photon at the microscopic level is predictable, yet observation of the current *direction* remains non-deterministic because we cannot know how many times a reversal takes place without disturbing the system. The absurd idea is that *reality is timeless* inside entangled systems (inspired by Barbour's timeless reality intuition [2]), i.e., it continually evolves and cycles through its recurrence, bound only by the available number of states. This symmetry can however be broken at the macroscopic level by an observer preparing the system for measurement, triggering a *direction* for the local flow of information, energy and causality.

P.L. Borrill (✉)
Earth Computing, Inc., Palo Alto, California, USA
e-mail: paul@borrill.com

A. Aguirre et al. (eds.), *It From Bit or Bit From It?*,
The Frontiers Collection, DOI 10.1007/978-3-319-12946-4_9

Subtime[1] (t_s) is introduced as a *reversible* information interchange within an entangled system. We re-examine a conclusion dismissed by Einstein, Podolsky & Rosen (EPR) [4]: we accept the principles of relativity and the constancy of the speed of light c (in t_s), but question our ability to *measure* c with experiments that *presume* a Classical Time (T_c); a smooth, monotonic and irreversible background [5] superimposed on a Minkowski spacetime manifold.

We propose an alternative view in the spirit of Boltzmann indistinguishability: in addition to the indiscernability of particles with identical properties [6] we recognize that states previously visited within a quantum system are *indistinguishable* from reversing *classical time* (T_c) to that prior state.

Information, Photons and Time

We begin by assuming that information is associated with Poynting's [7] propagation of a photon[2] and postulate that *subtime* is *inextricably intertwined* with *space* along the one-dimensional path bounded by the photon traversal between emitter and absorber atoms (a Shannon transmitter/receiver channel[3]).

We see no need for a four-dimensional (Minkowski) *background* for spacetime within which light cones are projected (in an empty manifold) to reason about causality, non-locality and the ordering of events.

In a nutshell, we dispense entirely with the notion that a background of time exists, along with any sense of future or past, *between* isolated entangled systems. Instead, reversible evolution *recurs* perpetually within an entangled system. Only when an entangled system decoheres into the environment of other entangled systems (through new photon exchanges) does time emerge as progressively irreversible, providing persistent evolution of information at the macroscopic scale.

Note that Feynman diagrams *implicitly* include a background of Minkowski spacetime. These diagrams we draw on our pieces of paper are not capable of depicting subtime (unless one were extraordinarily gifted in *origami*).

[1] Presented *without* mathematical description because existing formalisms contain implicit assumptions incompatible with this insight (in addition to their intrinsic Minkowski background, there are also two mutually incompatible forms of evolution–unitary/non-unitary). Einstein also believed the formalism was a hindrance to reasoning about quantum theory [3].

[2] Almost all Bell tests so far have been performed with photons [8]. This description may be applied to any quantum particle with a de Broglie wavelength; information simply travels at the slower rate of traversal of the particle through the apparatus. The helical path description is similar for electrons [9].

[3] Shannon [10] defined the notion of channel capacity in his theory of communication and the notion of a 'bit' as the fundamental unit of information.

The Absurd Idea

We propose a principle of *retroactive non-discernability* in the recurrence of states in entangled systems. Subtime paths (helicity eigenvalues) incur modular increments with photon traversals[4] from one atom to another and decrements on their return path resulting in a net zero change in subtime (t_s)[5] while (T_c) *appears to* stand still.

Instead of the assumed traversal of a photon (or other quantum particle) through a multiple-slit apparatus once only from the source to the detector, imagine a photon traversing backward and forward perpetually within the apparatus an arbitrary (uncountable) number of times before it is finally absorbed by an atom in the detector and passed on as an observation. We would be unable to detect (in any single measurement) how many traversals actually occurred before we registered the event in T_c. This implies:

- Most experimental observations would provide no clue that we were not measuring intervals in t_s. Instead we experience observation events in T_c like a quantum stroboscope, illuminating reality in *quick flashes* with long periods of *darkness* in between.
- Unlimited recurrence can take place within an entangled system in subtime. But (a) we would be unable to discern one recurrence from another from our T_c vantage point and (b) even for large systems of atoms many intermediate configuration states could be visited in their environment and then be reversed to a predecessor state before some external observation registered the state in T_c.
- All configurations may be explored in subtime; only those well suited to their environment would (with selective pressure) *persist* as (what would appear to be) *irreversible change* in T_c.

Entanglement and Recurrence

Time is change that we can count [15]. Two atoms exchanging a photon with each other *in perpetuity* comprise a bipartite entangled pair (Fig. 9.1). Each arrival of the photon (in t_s) at the atoms represents a gain in information and departure represents a loss, i.e., entropy. Information and subtime are incremented along the photon's path from the receiver's point of view and decremented from the point of view of the transmitter.

Each *entangled system* may evolve through its configuration space an arbitrary (and uncountable) number of times, but is inevitably constrained to a *recurrence* which is temporally indiscernible from any previous or successive recurrence.

---

[4] There are many theoretical and experimental investigations underway regarding the helical nature of photon propagation. Our contribution is recognizing that this is also a reversible action in *subtime*. Photons are also able to transfer multiple bits in higher order angular momentum [11–14].

[5] Consistent with the advanced and retarded wave solutions to Maxwell's equations.

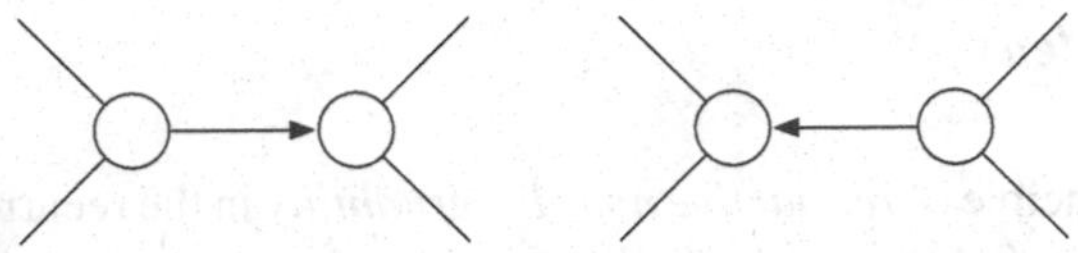

Fig. 9.1 A symmetric Heisenberg cut—photons can go both ways (a hot potato protocol in perpetuity)

Going from one to two atoms mediated by a photon, subtime becomes an isolated temporal experience of that two party system. As we add more atoms to the system, the number of discernible configurations increases non-linearly. The recurrence of the system becomes richer and more diverse, but the configuration space is still limited by the number of *retroactively discernible* configurations.

Every entangled system evolves independently or expands as it receives new energy and decays as it decoheres. We expect some power law distribution, e.g., the simplest two atom entangled system would be almost 100 % reversible in its state of perpetuity between recurrences. Progressively larger systems of atoms have both a larger space of recurrences as well as a smaller probability of reversing (de-evolving) to a previously visited state, simply because the number of states is so much larger. The *emergence* of irreversibility in T_c would rapidly approach 100 % as we observe larger and larger objects up the chain to our macroscopic world.

Information and Quantum Mechanics

Shannon information and Quantum Mechanics (QM) share a common context: probability. And all probability is conditioned on the actions of an observer, i.e., what binary (yes/no) questions the observer asks, either explicitly or implicitly. In QM, the minimum number of states (yes/no answers) needed to fully describe the system is exposed by the preparation of the measurement.
Our framework for this insight includes:

1. No common reference frame exists in empty spacetime.[6]
2. Space and (sub)time are inextricably intertwined in Poynting's revolving shaft along the path from transmitter to receiver[7] [7, 17]. Photons explore any number of bounded subtime elements, an indefinite number of times. This unitary evolution is computationally reversible.

[6] There is nothing in nature (or in any measurement carried out so far), supporting a *background of time*, which would allow us to discern temporal relationships between independent entangled systems. If a system has no interactions with other systems, there is no common frame of reference or coordinate system for time. Simultaneity, total and partial orders, are undefined.

[7] We take Feynmans clocks [16] literally.

3. Information is conserved in the photon link[8] between two atoms comprising an entangled system. Information transfer is negative with respect to the transmitter and positive with respect to the receiver. This symmetry is broken when an observation is prepared which triggers the flow of energy and information–establishing a casual and thermodynamic direction.
4. Causality is symmetric.[9] There is no privileged role or direction for the observer-observee relationship. For every action there is an equal and opposite reaction. Just as effects must have causes for them to exist, causes must also have effects for them to exist. Measurements of information will thus be different (and opposite in sign) for each observer from their vantage point.
5. Interactions are reversible. Links comprise a photon bouncing back and forth between a pair of atoms in a perpetual hot potato protocol. It is impossible to discern (in any individual measurement) the first traversal of information from A to B (or vice versa) from the $N+1$st traversal, i.e., N is fundamentally uncountable.
6. Many more events can occur in subtime (t_s) than can be observed from a T_c vantage point: well below any Nyquist threshold. Experiments will therefore yield random measurements of the quantum state to a T_c observer.

Information and Entanglement

Entanglement of quantum states is traditionally assumed to be a consequence of the principle of superposition. This phenomenon has confounded physicists since EPR [4] first drew our attention to its paradoxical nature. Insight to explain the experimental evidence that nature behaves quantum mechanically and non-locally has thus far been elusive.

EPR described two possible explanations for entanglement: (a) there was some interaction (simultaneous reality) between the particles despite their physical separation or (b) information about all possible outcomes was encoded in hidden variables. EPR preferred the second explanation because instantaneous action at a distance was in conflict with special relativity.

[8] Each link represents EPR's "simultaneous element of reality" [4]. Links are embedded in a quantum network automata, with each atom representing a vertex of bounded degree. This implies:

- A limit to the number of entanglement neighbors: partners with other nodes in the entangled system or with decoherence partners in the environment.
- Like the valency in atomic bonding, this implies that nature builds a multi-hop entanglement network out into the decoherence environment (similar to Figs. 9.3 and 9.4).
- Different particles may have different degrees. For example, photons have degree two: one transmitter and one receiver represents a Shannon channel.

[9] Solutions to the electromagnetic field equations are symmetric with respect to time inversion. This symmetry is reflected in all our fundamental laws of physics.

There is a third explanation: a flaw in the belief that time can be measured as a smooth, monotonically increasing point on a continuum.[10]

Time is change. When nothing changes, time stands still. When something changes, and then changes back, it is indistinguishable from time standing still.

Entanglement represents a state of *reversible change*; it is impossible to "count" (in an individual measurement) the number of recurrences within this state. This is one example of (apparent) randomness in quantum theory. It is not truly random (in the sense of being unpredictable). But it is *uncountable* because we cannot distinguish a single (one directional) exchange between two entities from any arbitrary odd number of exchanges; they are fundamentally indistinguishable in the T_c measurement events. The orthodox assumptions which may mislead us regarding a global background of time are:

The continuum assumption: The experience of an atom (receiver or transmitter of information) is *stroboscopic*; information change occurs abruptly at the instant (in t_s) of emission, or absorption of the photon by an atom. Although *motion* may be continuous (down to the Planck limit), it is the arrival of new information that presents a change of state in the receiver. These discontinuous events in t_s masquerade as a continuous flow in our underlying assumptions in T_c.

The irreversibility assumption: We assume from human experience [18] that time marches irreversibly forward. There is no evidence for this in physics. What we know is that if time (change) happens, we remember; if it happens and then the information reverses its path, we do not. Even behaviors that have already decohered in T_c which we might think to be immutable once they have *happened*, can (at least locally) *unhappen*, within the local T_c state record, along with our memories being reversed also [19].

An indefinite number of subtime units can be added and subtracted between the nodes in a quantum network, but only the net will be experienced by an observer. Different observers will also experience different measurements, because early observers will extract energy/information which will then be no longer available to other observers. Only a hypothetical witness with perfect *single traversal* properties could, in principle, detect the vector sum of subtime units in the system being measured. In practice, It is equally likely that the observer is fooled by the same stroboscopic, uncountable but indiscernible phenomena experienced by the system being measured.

Bell Experiments and Virtual Machines

Einstein proved that simultaneity was relative, but when we carry out Bell experiments, we set up our apparatus to detect coincidences with an implicit assumption that

[10] This recognition that the logic of the EPR paper was correct but the assumptions were wrong is shared by Nathan Rosen [3].

our observable measurements in T_c are equivalent to durations in t_s. Testing Bell's inequality requires two independent measurements (at points separated in space). Information regarding these measurements is signaled to a common site where coincidence is analyzed [20].

For the purpose of articulating this insight, imagine that *virtual machines*[11] (VMs) are used to carry out the experiments; one each at the separated points and a third at the common site to analyze the signals from the other two for coincidence. These VMs are governed by a clock cycle, orders of magnitude shorter than required to measure and analyze the results (equivalent to Aspect's atomic clock). In the spirit of Maxwell, imagine a demon,[12] which *suspends* and *resumes* each of the VMs (freezing them on a clock cycle) such that their periods of awareness do not overlap, but their computational state remains available while they are suspended and can be read by the others. The VMs have no independent timing reference, and have no idea that they are being time multiplexed in the t_{rt} (real-time) domain; their entire experience is governed by the events they observe in the T_{vm} (virtual machine) domain.

Now further imagine that these VMs are capable of reversible computation: the demon can allow the computation to proceed arbitrarily far into the algorithm, but at any point reverse that computation to some prior state visited by that VM. The equivalent of this, in the world of computing, is for the VM to be *reset* to some prior snapshot in order to re-acquire some previously consistent state. The VM has no idea it has been reset. Its only clue might be that its hardware time counters now differ from some external source of time that it may acquire from the network.

We can tune the rate of production of entangled photons such that they occur in the timing window of the measurement VMs, and the statistics of Bell states will emerge. However, this says nothing about what happens outside the timing windows, where any number of *internal* events my have taken place, i.e., any amount of forward or reverse computational evolution (non-Landauer [22] reversals or resets).

What is actually happening in the real world of Aspect's [23] Bell experiments? The apparent change in correlation (at distance) as soon as the polarizer is switched is explained simply by the reversal of subtime (the photon bouncing back), and a *rewriting of history* in T_c [24]. Which our instruments and memories would be *unable to remember* [19]—except perhaps for shadows left by the Pauli exclusion principle in the nearby atoms—an example where we can catch nature reversing itself even after we have made an observation. This implies that relativistic separability remains intact in t_s, while the temporal artifacts of violations in Bell's inequalities shows up in our T_c record.[13]

[11] Virtual machines in computing are software systems that emulate the hardware environment of a real computer, to allow one or many virtual machines (Operating Systems as well as applications) to run on the same physical hardware independently of one another.

[12] Aspect [21] measured time using randomly switched optical crystals at 50 MHz (20 ns), while the spatial extent of the apparatus required more than double that to violate special relativity. This is *not* what we are referring to as a demon.

[13] There is insufficient room in this paper to discuss distinctions with other "time loophole" theories. We draw the reader's attention to the principal arguments: that subtime starts and stops with the

From this insight, we can now begin to formulate a new and logically consistent *information* view of the apparent non-locality revealed in violations of Bell's inequality without sacrificing the principle of locality.

Information and Simultaneity

Since 1905 we often see assertions that—*there is no space without time*—because the speed of light provides a limit to the velocity of information traversal between atoms. We rarely hear the logically equivalent—*there is no time without space*—which is equally concludeable from Einstein's original postulates and argument [25]. Implications of this include:

- The notion of Minkowski space as a 4D manifold can mislead us that *time passes* independently of relative motion in space. We postulate that subtime does *not* flow when there is no motion along the path between emitter and absorber.
- Simultaneity surfaces, even in inertial frames, have no basis in reality. There is no common meaning to time separate from motion. They are inextricably tied together.
- Subtime intervals are EPR's "elements of simultaneous reality", terminated by the atoms on either end of the photon path. Subtime intervals are thus *finite*. The edges of the subtime graph are summed together to form the emergence of T_c. *Intervals* in time have been described by Barbour as an enigma: identified by Poincaré as an issue but otherwise remaining unresolved [26].
- The only *objective reality* that can be measured is through *interactions*—the ultimate *locality*. Entities must interact (touch, collide, bounce off, be absorbed, emitted etc.) in order to transfer information. However, the internal interactions of an entangled system are, by definition, unobservable. In T_c we observe only those rare events that touch the outside world through decoherence, below any Nyquist threshold.
- In bipartite entanglements, a photon (and its associated information) is trapped. It is perpetually bouncing between the atoms, just as virtual (photons) perpetually bounce between the orbiting electrons and protons in the nucleus of an atom.

We conclude that information is transmitted between atoms at a finite speed—the maximum being the speed of light—but question our ability to perceive this transmission as a *reversible* information-theoretic process. This creates an illusion of superluminal quantum-mechanical processes in experiments designed with a hidden assumption of a Minkowski spacetime background, which hinders our understanding of the EPR paradox.

(Footnote 13 continued)
emission and absorption of a photon, and is *reversed* in all ontological respects as the photon is returned in the hot-potato protocol. This is one way that we divorce ourselves from the background assumption of time, which is not (as far as we can tell) the case for other time loophole theories.

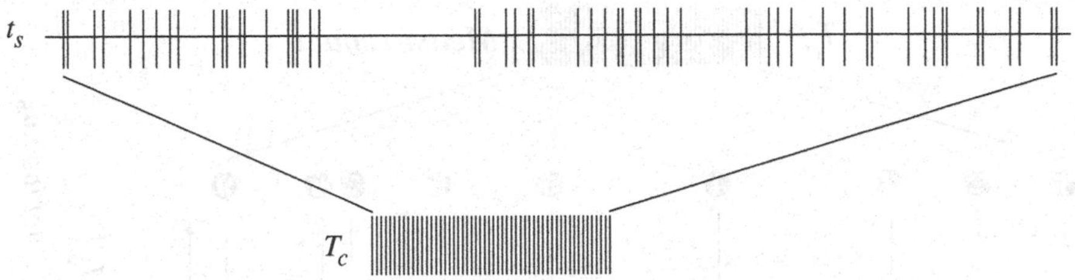

Fig. 9.2 Asynchronous events along an imaginary subtime line where events can arbitrarily interleave in a quantum network. *Subtime* (t_s) will *appear* continuous in *Classical Time* (T_c)

Subtime (t_s)

Subtime is what happens when we are not looking. It is the perpetual alternating direction of information flow through the bipartite interactions of atoms and photons. Subtime is (for our present argument) continuous, and is inseparable from the motion of photons (or other Boson).

Figure 9.2 shows asynchronous events along an imaginary t_s line, and the perception of these events back to back in T_C.

Subtime is reversible: everything that *happens* in subtime can *unhappen*. A photon that travels from A to B is *usually* followed by a traversal of that same photon from B to A. The state of the system is now indiscernible from that which existed before the first traversal, or indeed any prior or later traversal of the photon between them.[14]

We describe subtime as *propechronos*—from the Latin "propinquus"; (of space) near, neighboring, (of time) near, at hand, not far off; and "chronos"; the personification of time. To emphasize its locality (to the next atom in space), temporal symmetry, and mutual kinship with its bipartite entanglement partner.

Classical Time (T_c)

T_C *appears* successive, monotonic and irreversible and its sign is always positive (because it represents the absolute value of the sum of subtime intervals) in the network *trail*.

The perpetual *hot potato* photon[15] exchange in entanglement is timeless because we are unable to measure it with our instruments without taking energy out of the system (thus disturbing the state of entanglement).

[14] From the Lorentz frame of the photon, everything that happens inside the atom, between the absorption and its re-emission, will appear to have a proper time of Zero. The notion of instantaneous is a function of the arbitrary frame in which we chose to perform our calculations [3].

[15] Because photons are indistinguishable, photons in a perpetual hot potato protocol may compete with other photons *taking over* the entanglement [27]. Information and energy may, however, remain trapped within the same entangled system.

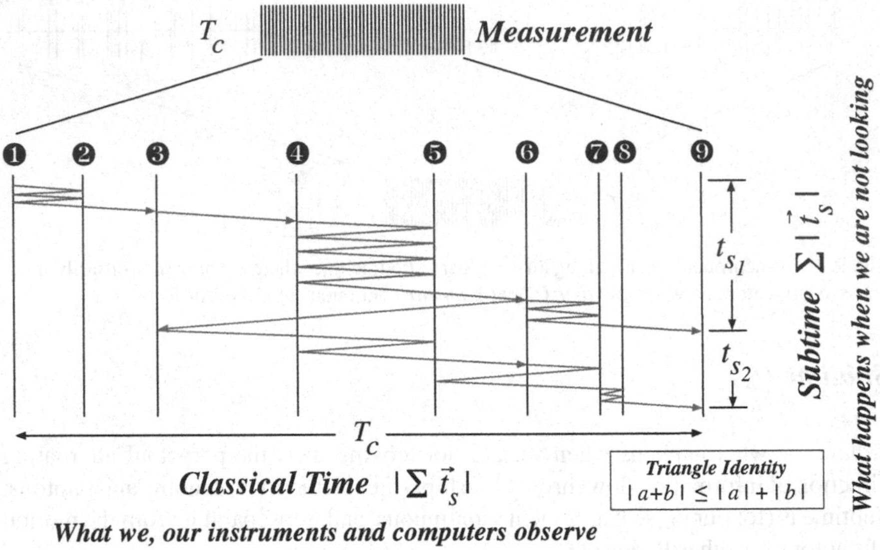

Fig. 9.3 Different accumulations in *Subtime* (t_s) can appear *the same* in Classical Time (T_c)

Figure 9.3 shows a photon traversing a chain of 9 atoms. The red path accumulates t_{s1} subtime units. The alternate green path (which continues half way through the red path before branching in a different direction) accumulates t_{s2} subtime units. Both will be experienced in T_c as the *same* interval of time. The order of events observed by different witnesses observing different atoms will therefore be different. This is in addition to the relativity of simultaneity in special relativity.

Extending the Entanglement Graph

Figure 9.4 shows a larger system of atoms, with three examples of alternate paths for the photon energy to travel between the energy/information ingress and egress points. The total length of the path defines t_s, but what we observe in T_c will be the absolute value of all the increments (forward traversals) in t_s, minus all the decrements (reverse traversals).

There are two segments of path ③ in Fig. 9.4 (second and fourth segments) illustrating entanglement as multiple photon reflections. Remember: an arbitrary odd number of reflections of this photon is indistinguishable from a single traversal. Also there is no way for us to discern (in an individual measurement) which path ①, ② or ③ was taken, and each has an arbitrary number of photon (subtime) reflections within them.

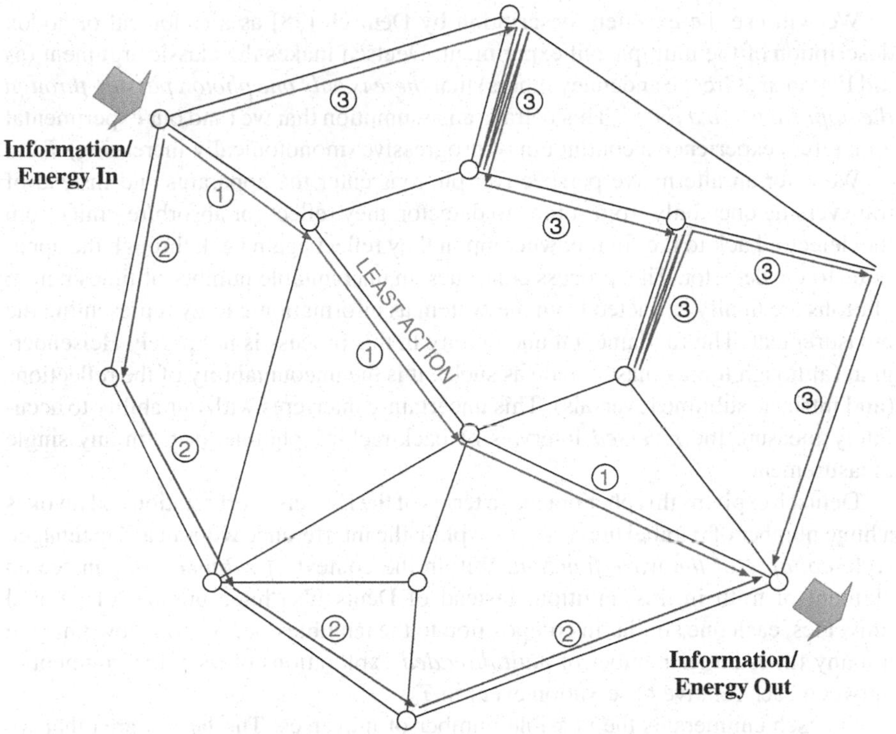

Fig. 9.4 A graph (2-D) view of different paths in an entangled system

- There is no distinction in the passage of time (in T_c) as far as the 'outside world' is concerned, with paths ①, ② or ③.
- Path ③ includes back and forth passing of the information/energy between the vertices in the graph. The number of passings *back and forth* is *uncountable*. This mixed path shows both temporary entanglement and direct *cut through* of photons through the system.
- There is no global passage of time. Each measurement experiences T_c as the *not yet reversed* receipt (and passing on) of information/energy within the entangled network.

Multiple Slit Experiment

It is commonly believed that if we decrease the intensity of a beam of light, we will eventually reach the point where only one photon is in transit through the apparatus *at a time*. An implicit but unacknowledged belief is that the photon travels only once (one way) through the apparatus from the intended source (transmitter) to the intended destination (receiver) whereupon energy (and information) is captured and the *measurement* is made.

We will use the excellent description by Deutsch [28] as a canonical orthodox description of the multiple slit experiment. Deutsch makes the classic argument (as did Feynman, Greene and many others) that *there is only one photon passing through the apparatus "at a time"*. This betrays an assumption that we (and our experimental apparatus) experience a continuum of progressive (monotonically increasing) T_c.

We offer an alternative perspective: photons enter the apparatus and instead of transversing once only from source to detector, they reflect (or absorb/re-emit) from the detector back to the source, whereupon they reflect again back through the apparatus to the detector. This process continues an uncountable number of times before photons are finally extracted from the system as information/energy representing the measurement. The fundamental uncertainty in this process is not purely Heisenbergian (although it may masquerade as such), it is the uncountability of the reflections (and intrinsic subtime reversals). This uncertainty interferes with our ability to accurately measure the *reversed* intervals of backtracking photons in t_s, in any single measurement.

Deutsch explains this phenomena in terms of the Everett interpretation and invokes a huge number of parallel universes to explain the interference without a Copenhagen style *collapse of the wave function*. Within the context of *subtime*, we can see an element of truth in this intuition. Instead of Deutsch's "huge number of parallel universes, each one similar in composition to the tangible one," we can now imagine a many times larger number of *multithreaded* explorations of its t_s environment—between each relative observation event in T_c.

Deutsch enumerates the possible number of universes. The largest area that we could conveniently illuminate with a laser might be about one square meter and the smallest manageable hole size might be a thousandth of a millimeter. So, there are approximately 10^{12} possible hole locations—alternative configurations—which can be explored in this system. It is critical to acknowledge that while this may be an approximate number for parallel universes, or with subtime exploring each location of the one square meter once, the entanglement in subtime expressed by our hypothesis is by its very nature *uncountable*, yet it exhibits a fundamental economy of mechanism and use of resources.

The implications of this include a *massive* unrealized concurrency under the hood of entangled information/subtime which is reminiscent of the hoped for *parallel computation* capacity of quantum computing.

We present a critical change in perspective: Instead of some magical parallel universes being explored, which is somehow beyond the relativistic physics of spatiality separated entities in some Bell-type inequality, we can now see ourselves and our instruments as observing the universe through a *time filter*. This time filter may be like a stroboscope or cinematographic projector. Each frame of the film represents a snapshot of subtime (events in t_s) and we can be fooled by our measurements into believing that there is zero "time" (T_c) between one frame to the next because all the change (the stroboscopic flash of reality) appears to occur at once. In our case, the step from one frame to another may be triggered asynchronously by individual decoherence events in T_c.

As the angle between the photon path(s) through the apparatus departs from the nominal 0°, we will observe interference through multiple slits as the phase of the helical photon path impinges on the target, with a *wave-like* probability of dark entanglement and light absorption/detection.

The geometry and mathematics of interference is well known but the mechanism traditionally used to explain it (waves) may now be compared with classical explanations within the *subtime* context. It is not that a single photon (or other quantum particle) is passing through both slits *at the same time*; it is passing back and forth with an indefinite number of traversals each reversing the effect in t_s. Appearing to traverse the apparatus only once in T_c because of our inability to accurately measure subtime intervals (between detection events in T_c), this *appears to* reinforce our assumption of a T_c background being smooth, monotonic and irreversible.

Entangled Systems Are *Dark*

An entangled system explores indefinitely within their *recurrences*, where the system neither gains nor loses energy/information. These are the maximally entangled states. Their existence will be *dark* i.e., *outside of time*—their existence is unobservable in T_c, either as *emitters* or *absorbers*. The answer to the question, "Is the moon there when we are not looking?" is *yes*; however, it hides from us in *subtime* (this should not be taken literally).

In t_s photons may take any and all paths that exist in the apparatus an uncountable number of times between each detection event. Between *detection events* photons are trapped/hidden, thus perpetuating the *state of darkness*. This provides a potentially straightforward (classical) explanation for: interference in the two-slit experiment, Feynman's glass reflector system, quantum erasure, quantum teleportation, the quantum zeno effect and entanglement swapping [27].

Falsifiability

Many experiments can be conceived to prove this conjecture incorrect. Below are a small sample of how it may be tested experimentally:

- Separate (non-interacting) entangled systems will develop (evolve their state within the constraints of recurrence) entirely independently. No background of time exists which is common to all systems in even our local world. Independent atomic clocks will exhibit random (unexplained) perturbations relative to each other. These jumps will be affected by an increase in electromagnetic coupling to other systems (onset of decoherence).
- The hoped-for parallelism in quantum computing *may not* be achievable for reasons more fundamental than the practical difficulties with decoherence. For

example, benchmark results of the first 512-bit quantum computer snowed no quantum speedup relative to a standard desktop computer [29].

These results are far from conclusive. However, the subtime conjecture is a potential new avenue of investigation that could lead to new insights, and to experimentally verifiable properties. In principle, the benefits of massive distributed concurrency, entanglement, state teleportation and other quantum phenomena may be achievable through mechanisms emulated by networks of computers.

Conclusions

Photons are the carrier of time and the Universe is a *network automaton*[16]: a graph of evolving relationships where the vertices represent atoms and the edges represent the hot-potato protocol of a continuously (in perpetuity) bouncing back and forth of a photon. The concept of subtime carries many of the hallmarks of *entanglement*.

Photon entanglements represent reversible, bounded *intervals* of reversible subtime. Indeed, the only realistic intervals that we can measure are those that span the space/time path of the photon and are *terminated* by the atoms. Intervals in subtime are therefore finite and bounded by the symmetric emitter and absorber atoms [30].

What goes on inside entangled systems is both *timeless* and *unobservable*. Only *rare* interactions (observations) with the outside define the order of events that we see. Entangled systems are *dark*.

Many more events can occur in subtime (t_s) than can be observed from a T_c vantage point: well below any Nyquist threshold. Quantum measurements will thus yield random results.

We question the idea that massive concurrency exists in quantum computation, and suggest instead that we have been sampling subtime like a stroboscope in T_c: we see brief flashes of reality with long periods of darkness in between. We also recognize the intuition behind multiple parallel universes. Instead we imagine entangled systems to exhibit unbounded exploration of the quantum state space in t_s, not dissimilar to a conventional computer multithreading with[17] many tasks vying for its physical resources and our apparent *random* selection of the current state of one of the threads through a deliberate (or otherwise) preparation of our observations.

> We must, therefore, be prepared to find that further advance into this region will require a still more extensive renunciation of features which we are accustomed to demand of the space time mode of description.
>
> –*Niels Bohr* [30]

Since the publication of an earlier version of this article (In July 2013), experimental verification of the emergence of time from quantum entanglement has been illustrated [31].

---

[16] A network automaton is similar to a cellular automaton, but where the cells are vertices in an arbitrary network and there is an evolving topology of links connecting them.

[17] Multithreading is distinguished from multiprocessing in computer systems, in that *threads* share the resources of a single computer. i.e., we have one universe, not many.

Acknowledgments These ideas were inspired by the writings and conversations with Lee Smolin, Julian Barbour, Fotini Markopoulou, Simone Severini and Anton Zeilinger. They may not agree with anything I have said, but I owe my inspiration to the questions they asked. All responsibility for errors and inaccuracies is mine.

References

1. C. Teuscher (ed.), *Alan Turing: Life and Legacy of a Great Thinker* (Springer, New York, 2004)
2. J. Barbour, *The End of Time: The Next Revolution in Physics* (Oxford University Press, 2001)
3. A. Peres, "Einstein, Podolsky, Rosen, and Shannon", Technion-Israel Institute of Technology, October 2003. arXiv e-print arXiv:quant-ph/0310010
4. A. Einstein, B. Podolsky, N. Rosen, Can quantum-mechanical description of physical reality be considered complete? Phys. Rev. **47**, 777–780 (1935)
5. L. Smolin, "The case for background independence", Perimeter Institute, July 2005. arXiv e-print arXiv:hep-th/0507235
6. S. French, D. Krause, *Identity in Physics: A Historical, Philosophical, and Formal Analysis* (Oxford University Press, New York, 2006)
7. J.H. Poynting, The wave motion of a revolving shaft, and a suggestion as to the angular momentum in a beam of circularly polarised light. Proc. R. Soc. Lond. Ser. A **82**, 560–567 (1909)
8. A. Shimony, Aspects of nonlocality in quantum mechanics, in *Quantum Mechanics at the Crossroads New Perspectives from History, Philosophy and Physics*, ed. by J. Evans, A.S. Thorndike (Springer, Berlin, 2010)
9. D. Hestenes, "Electron time, mass and Zitter", technical report, FQXi Community (2008)
10. C.E. Shannon, A mathematical theory of communication. Bell Syst. Tech. J. **5**(1), 55 (1948)
11. R. Fickler, R. Lapkiewicz, W.N. Plick, M. Krenn, C. Schaeff, S. Ramelow, A. Zeilinger, Quantum entanglement of very high angular momenta. Science **338**, 640–643 (2012). University of Vienna, July 2012. arXiv e-print arXiv:1207.2376
12. J.P. Torres, L. Torner, *Twisted Photons: Applications of Light with Orbital Angular Momentum* (Wiley, Weinheim, 2011)
13. A. Afanasev, C.E. Carlson, A. Mukherjee, Excitation of an atom by twisted photons, George Washington University, March 2013. arXiv e-print arXiv:1304.0115 [quant-ph]
14. J. Bahrdt, K. Holldack, P. Kuske, R. Mller, M. Scheer, P. Schmid, First observation of photons carrying orbital angular momentum in undulator radiation. Phys. Rev. Lett. **111**, 034801 (2013)
15. U. Coope, *Time for Aristotle* (Oxford University Press, Oxford, 2005)
16. R.P. Feynman, A. Zee, *QED: The Strange Theory of Light and Matter* (Princeton University Press, Princeton, 2006)
17. J.H. Poynting, On the transfer of energy in the electromagnetic field. Philos. Trans. R. Soc. Lond. **175**, 343–361 (1884)
18. D. Gentner, M. Imai, L. Boroditsky, As time goes by: evidence for two systems in processing space time metaphors. Lang. Cognit. Process. **17**(5), 537565 (2002)
19. L. Maccone, Quantum solution to the arrow-of-time dilemma. Phys. Rev. Lett. **103**(8), 080401 (2009)
20. A. Peres, Quantum nonlocality and inseparability, in *New Developments on Fundamental Problems in Quantum Physics*, ed. by M. Ferrero, A. van der Merwe (Kluwer, Dordrecht, 1997), pp. 301–310. arXiv e-print arXiv:quant-ph/9609016, Technion-Israel Institute of Technology, September 1996
21. A. Aspect, J. Dalibard, G. Roger, Experimental test of bell's inequalities using time-varying analyzers. Phys. Rev. Lett. **49**, 1804–1807 (1982)
22. C.H. Bennett, R. Landauer, The fundamental physical limits of computation. Sci. Am. **253**(1), 48–56 (1985)

23. A. Aspect, Bell's inequality test: more ideal than ever. Nature **398**(6724), 189–189 (1999)
24. C.H. Bennett, Notes on Landauer's principle, reversible computation and Maxwell's demon. Stud. Hist. Philos. Mod. Phys. **34**, 501–510 (2003). arXiv:physics/0210005, October 2002
25. A. Einstein, On the electrodynamics of moving bodies (zur elektrodynamik bewegter korper). Annalen der Physik **XVII**, 891–921 (1905)
26. J. Barbour, The nature of time, technical report, FQXi Community, Mar (2009)
27. E. Megidish, A. Halevy, T. Shacham, T. Dvir, L. Dovrat, H.S. Eisenberg, Entanglement swapping between photons that have never coexisted. Phys. Rev. Lett. **110**, 210403 (2013)
28. D. Deutsch, *The Fabric of Reality: The Science of Parallel Universes- and Its Implications* (Penguin Books, New York, 1998)
29. W. Vinci, T. Albash, A. Mishra, P.A. Warburton, D.A. Lidar, "Distinguishing classical and quantum models for the d-wave device". arXiv:1403.4228 [quant-ph] March 2014
30. J.A. Wheeler, R.P. Feynman, Interaction with the absorber as the mechanism of radiation. Rev. Mod. Phys. **17**, 157–181 (1945)
31. M. Ekaterina, G. Brida, M. Gramegna, V. Giovannetti, L. Maccone, M. Genovese. "Time from Quantum Entanglement: An Experimental Illustration." Phys. Rev. A **89**(5), (2014). arxiv.org/abs/1310.4691, October 2013

Chapter 10
These from Bits

Yutaka Shikano

Operational Derivation of Physical Laws

When answering the question of what properties a material has, a theoretical physicist may ask

> "What is its Hamiltonian?" or "What is its Lagrangian?"

Most physicists seem to believe that every physical property of a material can be predicted once the Hamiltonian or Lagrangian of some physical phenomena are known. This is often called "physics imperialism." In the 20th century, we perhaps benefited too much from practical developments in physics—semiconductors, lasers, and magnetometers. To reinforce its position, the 20th century saw physics expanding the boundaries of various physical phenomena, from the sub-nanometer to the cosmological scale.

On the other hand, when somebody asks the same question to a non-expert physicist, they may try to break open the object with a hammer, for example, or measure its electrical properties. That is, to reveal the attributes of this material, they take a step-by-step approach. We can think of this as *operational* thinking. This method is very powerful when it comes to understanding unknown physical phenomena. Further, operational thinking is a natural process for all experimentalists. To reveal a material's physical properties, experimentalists construct their experimental setup, start the detection by flicking a switch, measure something, and then analyze the experimental data. Obviously, before the experimental setup has been constructed, we cannot collect experimental data. This is essentially a step-by-step (operational)

Y. Shikano (✉)
Research Center of Integrative Molecular Systems (CIMoS),
Institute for Molecular Science, Okazaki 444-8585, Japan
e-mail: yshikano@ims.ac.jp

Y. Shikano
Institute for Quantum Studies, Chapman University,
Orange, CA 92866, USA

A. Aguirre et al. (eds.), *It From Bit or Bit From It?*,
The Frontiers Collection, DOI 10.1007/978-3-319-12946-4_10

process. In order to naturally understand physical properties via such a process, it seems to be necessary to reconstruct all physical laws from an operational point of view.

Operational thinking has been formalized as information theory. Historically, as recounted in the book "Science and Information Theory," Leon Brillouin tried to apply this theory to physical laws [1]. His book aims to capture various physical phenomena from the information-theoretical idea initiated by Claude Elwood Shannon. In particular, he tried to derive the entropy of physical systems from the information-theoretical quantity known as the Shannon entropy. From an information theory standpoint, the Shannon entropy can be thought of as the averaged rate of the optimal data compression [2]. This seems to fit the concept of John Archibald Wheeler's famous quote:

"It from Bit."

However, as shown in the next section, information-theoretical concepts cannot be applied to a single event. In this essay, we show that this quote should in fact be rewritten as:

"These from Bits."

Individuals and Information Theory

First of all, how should we evaluate the quantity of information? For example, the abbreviation "IMS"[1] has the following ASCII binary code:

IMS ⇒ 010010010100110101010011

Thus, "IMS" has a 24-bit string. However, nobody would claim that the Shannon entropy of "IMS" is 24. Furthermore, the 24-bit string alone has no meaning. For example, another abbreviation, "MIT," can be converted to

MIT ⇒ 010011010100100101010100

This also has a 24-bit string, but the meaning of the two abbreviations is completely different. Therefore, the amount of information does not reflect the meaning of each word. So what does the amount of information express? Neither word has an information-theoretical meaning. Therefore, we have to define the amount of information for an ensemble of bit strings. For example, we could consider the set of bit strings given by "CIT," "NIT," and "TSU," and evaluate the probability distribution of the bit string pattern, e.g., the ratio of the number of 1's. However, this probability distribution cannot be evaluated from just a single event. Therefore, we require an ensemble containing a large number of samples. Then, for a sufficiently large number of samples, the probability distribution becomes the "true" probability distribution. In this case, each bit string is called a typical sequence.

---

[1] "IMS" stands for "Institute for Molecular Science," which is the author's working institute.

For a typical sequence of N bits, Shannon analytically showed that the optimal data compression rate could be written as

$$\tilde{N} = NH(p), \tag{10.1}$$

where $\tilde{N}$ is the averaged number of the optimally compressed bits,[2] and $H(p)$ is the Shannon information for the bit string, which is given by

$$H(p) = -p \log_2 p - (1-p) \log_2 (1-p), \tag{10.2}$$

where p is the ratio of the number of 1's in the bit string. Therefore, on applying information theory to physical laws, macroscopic systems, such as those of thermodynamics and statistical mechanics, are needed. Information theory cannot be applied to Newtonian mechanics and electromagnetism, as the theory breaks down for small data sets or a single event. However, in our physical experiences and daily life, such phenomena or events are commonly encountered. We must therefore construct a relevant description of information theory on this scale.

Equilibrium Thermodynamics and Statistical Mechanics from an Operational Viewpoint

In the previous section, we showed that information theory can only be applied to physical systems with a macroscopically large number of samples. As is well known, the macroscopic theory of physics is described by thermodynamics and statistical mechanics. Let us first consider the structure of thermodynamics. Equilibrium thermodynamics itself has an operational perspective, and, further, it can be axiomatized by a specific operational process, namely the adiabatic process [3].[3] Therefore, the long history of thermodynamics can be placed into an information-theoretical context. The famous parallel between thermodynamics and information theory is the paradox of Maxwell's demon [4], explained as follows. Consider a molecular gas inside a box. The box contains a partition that divides it into two regions, and the partition has a window that can be either open or shut. The demon operates this window. When the demon sees molecules moving at higher speeds, he guides them to the left side of the box via the window. Similarly, the demon guides molecules moving at lower speeds to the right side of the box. The demon repeats this process repeatedly. Eventually, the temperature in the left of the box increases, and vice versa. This seems to violate the second law of thermodynamics, and was taken as the paradoxical issue. However, Rolf William Landauer pointed out that the

---

[2] Shannon originally showed that there exists some lower bound of the (reversible) compression process such that $NH(p) \le \tilde{N} < NH(p) + 1$ for any N-bit string.

[3] The same authors recently showed that nonequilibrium thermodynamics cannot, in general, be defined in the same way [10].

mind of the demon retains the memory of the molecular speed, and further that the erasure of this memory must incur some cost [5]. This cost is equivalent to the gain from the physical system. Therefore, by considering not only the thermodynamical cycle but also the information cycle, the second law of thermodynamics is not violated. Further developments on the resolution of the Maxwell's demon paradox have been contributed by various researchers, particularly Charles Henry Bennett [6, 7]. However, there remains an unsolved problem of the relationship between the thermodynamical entropy of the physical system and the Shannon entropy of the demon. In Ref. [8], we pointed out the equivalence between these entropies when the cleverest Maxwell's demon operates the physical and information-theoretical processes in a specific context. These physical processes do not incur any cost from the operation of the partition, the window, or the measurement. We can also ensure that the information-theoretical processes do not incur any computational cost in the demon's memory. Only when the cleverest Maxwell's demon applies the optimal data compression to his memory before the erasure does the Shannon entropy equal the thermodynamical entropy. Therefore, if all molecules in the box are measured by the cleverest demon, the thermodynamical entropy in all of the thermodynamical processes can be characterized by the Shannon entropy in the information-theoretical context. Hence, "These (thermodynamical processes) from Bits."

Next, let us consider another macroscopic physical theory: statistical mechanics. In equilibrium statistical mechanics, we conventionally discuss a derivation of the ground state of a sufficiently large number of spins and a phase transition from liquid to solid, for example. Equilibrium statistical mechanics does not have an operational structure. Therefore, to pursue our idea that any physical process can be reformulated from an operational viewpoint, we must construct some operational scenarios. For simplicity, consider a physical system with N two-level atoms. Somebody, who we symbolize as Maxwell's demon in the following, measures each two-level atom. First, Maxwell's demon measures the N-ary physical system. The demon's memory stores the bit-string of the excited state (1) or the ground state (0), and so the demon incurs the optimal erasure cost[4] given by

$$W_{era}(p) = N\, H(p)\, k_B T \ln 2 \tag{10.3}$$

where p denotes the ratio of the number of excited states, k_B is the Boltzmann constant, and T is the temperature of the heat bath in the physical erasure model. From Landauer's well-known principle, the averaged cost of the erasure process is $k_B T \ln 2$. We can also determine the cost of exciting the physical system from the ground state for all two-level atoms as

$$W_{phys}(p) = N p\, \epsilon \tag{10.4}$$

[4] We consider the optimal erasure cost because equilibrium thermodynamics can be equated to equilibrium statistical mechanics.

for the two-level energy difference ϵ. Then, we define the cost function $F(p)$ as

$$F(p) := W_{phys}(p) - W_{era}(p). \tag{10.5}$$

Intuitively, one of the essential properties of the equilibrium state is its robustness against small perturbations to the physical system. In our operational context, we define the equilibrium state as the robustness of the cost function $F(p)$ under a small change to the physical system:

$$\frac{dF(p)}{dp} = 0 \tag{10.6}$$

for sufficiently large N [9]. Thus, we can derive the Maxwell–Boltzmann distribution as

$$\frac{\text{the number of } 1's}{\text{the number of } 0's} = \frac{p}{1-p} = \exp\left(-\frac{\epsilon}{k_B T}\right). \tag{10.7}$$

To conclude, we derive the Maxwell–Boltzmann distribution, which is the conventional derivation of the equilibrium state in statistical mechanics from an operational statistical process with optimal data compression and erasure processes.[5] Once again, therefore, we have "These (physical systems to satisfy statistical physics) from Bits."

Concluding Remarks

Following in Brillouin's footsteps, we tried to reformulate some physical theories from an operational viewpoint. However, as information theory is not currently applicable to situations where there are only a small number of samples, we could only consider macroscopic physical theories: equilibrium thermodynamics and equilibrium statistical mechanics. The optimal information-theoretical process corresponds to the equilibrium macroscopic system, and its essence is a sufficiently large number of samples. Therefore, Wheeler's famous slogan should be changed to "These from bits." To revive the original "It from bit," we must extend information theory to small-number samples or non-typical sequences. I believe that microscopic physical theories, such as Newtonian mechanics, can play a great part in the development of information theory. At such a time, "It develops Bit," and we will surely acquire "It from Bit."

[5] Our approach is completely different from that of Jaynes [11], as seen in Ref. [9, Appendix B].

References

1. L. Brillouin, *Science and Information Theory* (Dover, New York, [1956, 1962] 2004)
2. C.E. Shannon, Bell Syst. Tech. J. **27**, 379 (1948), 623 (1948)
3. E.H. Lieb, J. Yngvason, Phys. Rep. **310**, 1 (1999)
4. J.C. Maxwell, *Theory of Heat* (Longmans, Green, London, 1871), pp. 308–309
5. R. Landauer, IBM J. Res. Dev. **5**, 183 (1961)
6. C.H. Bennett, IBM J. Res. Dev. **17**, 525 (1973)
7. C.H. Bennett, Int. J. Theor. Phys. **21**, 905 (1982)
8. A. Hosoya, K. Maruyama, Y. Shikano, Phys. Rev. E **84**, 061117 (2011)
9. A. Hosoya, K. Maruyama, Y. Shikano, arXiv:1301.4854
10. E.H. Lieb, J. Yngvason, Proc. R. Soc. A **469**, 20130408 (2013)
11. E.T. Jaynes, Phys. Rev. **106**, 620 (1957)

Chapter 11
Self-similarity, Conservation of Entropy/bits and the Black Hole Information Puzzle

Douglas Singleton, Elias C. Vagenas and Tao Zhu

Abstract John Wheeler coined the phrase "it from bit" or "bit from it" in the 1980s. However, much of the interest in the connection between information, i.e. "bits", and physical objects, i.e. "its", stems from the discovery that black holes have characteristics of thermodynamic systems having entropies and temperatures. This insight led to the information loss problem—what happens to the "bits" when the black hole has evaporated away due to the energy loss from Hawking radiation? In this essay we speculate on a radical answer to this question using the assumption of self-similarity of quantum correction to the gravitational action and the requirement that the quantum corrected entropy be well behaved in the limit when the black hole mass goes to zero.

Published in *JHEP* **1405**:074 (2014).

D. Singleton (✉)
Department of Physics, California State University Fresno,
93740-8031 Fresno, CA, USA
e-mail: dougs@csufresno.edu

D. Singleton
Department of Physics, Institut Teknologi Bandung,
Jalan Ganesha 10, Bandung 40132, Indonesia

E. C. Vagenas
Theoretical Physics Group, Department of Physics,
Kuwait University, P.O. Box 5969, 13060 Safat, Kuwait
e-mail: elias.vagenas@ku.edu.kw

T. Zhu
GCAP-CASPER, Physics Department, Baylor University,
76798-7316 Waco, TX, USA
e-mail: Tao_Zhu@baylor.edu

T. Zhu
Institute for Advanced Physics and Mathematics,
Zhejiang University of Technology, 310032 Hangzhou, China

A. Aguirre et al. (eds.), *It From Bit or Bit From It?*,
The Frontiers Collection, DOI 10.1007/978-3-319-12946-4_11

Self-Similarity and Order-$\hbar^n$ Quantum Gravity Corrections

In this essay we look at the connection between physical objects, i.e. "its", and information/entropy, i.e. "bits",[1] in the context of black hole physics. In particular, we focus on the relationship between the initial information/entropy contained in the horizon of a Schwarzschild black hole and the final entropy carried by the outgoing, *correlated* photons of Hawking radiation. The correlation of the photons comes from taking into account conservation of energy and the back reaction of the radiation on the structure of the Schwarzschild space-time in the tunneling picture [2, 3] of Hawking radiation. Since, in the first approximation, Hawking radiation is thermal there are no correlations between the outgoing Hawking radiated photons. This leads to the information loss puzzle of black holes which can be put as follows: The original black hole has an entropy given by $S_{BH} = \frac{4\pi k_B G M^2}{c\hbar}$ which can be written as $S_{BH} = \frac{k_B A}{4 l_{Pl}^2}$ where $A = 4\pi r_H^2$ is the horizon area of the black hole and $r_H = \frac{2GM}{c^2}$ is the location of the horizon [4]. One can think of this areal entropy as being composed of Planck sized area "bits", $A_{Pl} = l_{Pl}^2$, where the Planck length is defined as $l_{Pl} = \sqrt{\frac{\hbar G}{c^3}}$. If Hawking radiation were truly thermal, then the entropy of the outgoing thermal radiation would be larger than this Bekenstein area entropy. Since entropy increases, some information is lost. But this violates the prime directive of quantum mechanics that quantum evolution should be unitary and, thus, information and entropy should be conserved.

To begin our examination of these issues of the thermodynamics of black holes and the loss versus conservation of information, we lay out our basic framework. We will consider a massless scalar field $\phi(\mathbf{x}, t)$ in the background of a Schwarzschild black hole whose metric is given by

$$ds^2 = -\left(1 - \frac{2M}{r}\right) dt^2 + \frac{1}{\left(1 - \frac{2M}{r}\right)} dr^2 + r^2 d\Omega^2, \tag{11.1}$$

in units with $G = c = 1$. From here onward in the essay we will set $G = c = 1$ but will keep $\hbar$ explicitly. The horizon is located by setting $1 - \frac{2M}{r_H} = 0$ or $r_H = 2M$. Into this space-time, we place a massless scalar field obeying the Klein-Gordon equation

$$-\frac{\hbar^2}{\sqrt{-g}} \partial_\mu (g^{\mu\nu} \sqrt{-g} \partial_\nu) \phi = 0. \tag{11.2}$$

By the radial symmetry of the Schwarzschild space-time as given by Eq. (11.1), the scalar field only depends on r and t. Expanding $\phi(r, t)$ in a WKB form gives

[1] There is an equivalence or connection between information, entropy and bits and we will use these terms somewhat interchangeably throughout this essay. A nice overview of the close relationship between information, entropy and bits can be found in reference [1].

$$\phi(r,t) = \exp\left[\frac{i}{\hbar} I(r,t)\right] \tag{11.3}$$

where $I(r, t)$ is the one-particle action which can be expanded in powers of $\hbar$ via the general expression

$$I(r,t) = I_0(r,t) + \sum_{j=1}^{\infty} \hbar^j I_j(r,t). \tag{11.4}$$

Here, $I_0(r, t)$ is the classical action and $I_j(r, t)$ are order $\hbar^j$ quantum corrections. We now make the assumption that quantum gravity is *self-similar*[2] in the following sense: the higher order corrections to the action, $I_j(r, t)$, are proportional to $I_0(r, t)$, i.e. $I_j(r, t) = \gamma_j I_0(r, t)$ where γ_j are constants. With this assumption, Eq. (11.4) becomes

$$I(r,t) = \left(1 + \sum_{j=1}^{\infty} \gamma_j \hbar^j\right) I_0(r,t). \tag{11.5}$$

From Eq. (11.5), one sees that $\gamma_j \hbar^j$ is dimensionless. In the units we are using, i.e. $G = c = 1$, $\hbar$ has units of the Planck length squared, i.e. l_{Pl}^2, thus γ_j should have units of an inverse distance squared to the jth power. The natural distance scale defined by Eq. (11.1) is the horizon distance $r_H = 2M$, thus

$$\gamma_j = \frac{\alpha_j}{r_H^{2j}} \tag{11.6}$$

with α_j dimensionless constants which we will fix via the *requirement* that information/entropy be well behaved in the $M \to 0$ limit. Thus, in this way we will obtain an explicit, all orders in $\hbar$ correction to the entropy and show how this gives a potential solution to the black hole information puzzle.

Black Hole Entropy to All Orders in $\hbar$

In [6] the set-up of the previous section was used to obtain an expression for the quantum corrected temperature of Hawking radiation [7] to all orders in $\hbar$. This was done by applying the tunneling method introduced in [2, 3] to the WKB-like expression given by Eqs. (11.3), (11.5) and (11.6). From [6], the quantum corrected Hawking temperature is given as

---

[2] Broadly speaking, self-similarity means that a system "looks the same" at different scales. A standard example is the Koch snowflake [5] where any small segment of the curve has the same shape as a larger segment. Here, self-similarity is applied in the sense that as one goes to smaller distance scales/higher energy scales by going to successive orders in $\hbar$ that the form of the quantum corrections remains the same.

$$T = \frac{\hbar}{8\pi M}\left(1 + \sum_{j=1}^{\infty} \frac{\alpha_j \hbar^j}{r_H^{2j}}\right)^{-1} . \tag{11.7}$$

In this expression, $\frac{\hbar}{8\pi M}$ is the semi-classical Hawking temperature and the other terms are higher order quantum corrections. At this point, since the α_j's are completely undetermined, the expression in Eq. (11.7) does not have much physical content but is simply a parameterizing of the quantum corrections. However, by requiring that the quantum corrected black hole entropy be well behaved in the limit $M \to 0$, we will fix α_j's and show how this leads to conservation of information/entropy, thus providing an answer to the black hole information loss puzzle.

Using Eq. (11.7), we can calculate the Bekenstein entropy to all orders in $\hbar$. In particular, the Bekenstein entropy of black holes can be obtained by integrating the first law of thermodynamics, $dM = TdS$ with the temperature T given by Eq. (11.7), i.e. $S = \int \frac{dM}{T}$. Integrating this over the mass, M, of the black hole (and recalling that $r_H = 2M$) gives the modified entropy as a function of M

$$S_{BH}(M) = \frac{4\pi}{\hbar} M^2 + \pi\alpha_1 \ln\left(\frac{M^2}{\hbar}\right) - \pi \sum_{j=1}^{\infty} \frac{\alpha_{j+1}}{4^j j}\left(\frac{\hbar}{M^2}\right)^j . \tag{11.8}$$

To lowest order $S_0(M) = \frac{4\pi}{\hbar} M^2$ for which the limit $M \to 0$ is well behaved, i.e. $S_0(M \to 0) \to 0$, as expected since as the mass vanishes so should the entropy. On the other hand, for the first, logarithmic correction as well as the other higher corrections, the quantum corrected entropy diverges. One way to fix these logarithmic and power divergences in $S_{BH}(M)$ as $M \to 0$ is to postulate that the Hawking radiation and resulting evaporation turn off when the black hole reaches some small, "remnant" mass m_R [8]. Here, we take a different path—by assuming that quantum corrected black hole entropy should not diverge in the $M \to 0$ limit we will obtain a condition that fixes almost all the unknown α_j's. To accomplish this, the third term in Eq. (11.8) should sum up to a logarithm which can then be combined with the second logarithmic term to give a non-divergent entropy, i.e. $S(M \to 0) \neq \pm\infty$. This condition can be achieved by taking the α_j's as

$$\alpha_{j+1} = \alpha_1 (-4)^j \quad \text{for} \quad j = 1, 2, 3 \ldots . \tag{11.9}$$

This again shows self-similarity since all the α_j's are proportional to each other. For this choice in Eq. (11.9), the sum in Eq. (11.8), i.e. the third term, becomes $+\alpha_1 \pi \ln(1 + \hbar/M^2)$. Combining this term with the second, logarithmic quantum correction, the entropy takes the form

$$S_{BH}(M) = \frac{4\pi}{\hbar} M^2 + \pi\alpha_1 \ln\left(1 + \frac{M^2}{\hbar}\right) . \tag{11.10}$$

As $M \to 0$, this "all orders in $\hbar$" entropy tends to zero, i.e. $S_{BH}(M) \to 0$. There is a subtle issue with identifying the sum in Eq. (11.8) with $\alpha_1 \pi \ln(1 + \hbar/M^2)$—strictly this is only valid for $\sqrt{\hbar} < M$, i.e. when the mass, M, is larger than the Planck mass. However, we can use analytic continuation to define the sum via $\alpha_1 \pi \ln(1 + \hbar/M^2)$ even for $\sqrt{\hbar} > M$. This is analogous to the trick in String Theory [14] where the sum $\sum_{j=1}^{\infty} j$ is defined as $\zeta(-1) = -1/12$ using analytic continuation of the zeta function, i.e. $\zeta(s) = \sum_{n=1}^{\infty} n^{-s}$. Other works [11–13] have investigated quantum corrections to the entropy beyond the classical level. These expressions, in general, involve logarithmic and higher order divergences as $M \to 0$ as we also find to be the case for our generic expression in Eq. (11.8). However, here, as a result of our assumption of self-similarity of the $\hbar^n$ corrections, we find an expression for $S_{BH}(M)$ which has a well behaved $M \to 0$ limit.

This "lucky" choice of α_j's in Eq. (11.9) which gave the all orders in $\hbar$ expression for $S_{BH}(M)$ in Eq. (11.10) was motivated by making the primary physical requirement that the entropy of the black hole be well behaved and finite. Usually, the focus in black hole physics is to find some way to tame the divergent Hawking temperature in the $M \to 0$ limit whereas here the primary physical requirement has been on making sure that the entropy/information content of the black hole is well behaved to all orders in $\hbar$.

The expression for $S_{BH}(M)$ still contains an arbitrary constant, namely α_1, which is the first order quantum correction. This first order correction has been calculated in some theories of quantum gravity. For example, in Loop Quantum Gravity one finds that $\alpha_1 = -1/2$ [15]. Once α_1 is known, our assumption of self-similarity and the requirement that information/entropy be well behaved fixes the second and higher order quantum corrections. One can ask how unique is the choice in Eq. (11.9)? Are there other choices which would yield $S_{BH}(M = 0) \to 0$? As far as we have been able to determine, there are no other choices of α_j's that give $S(M = 0) \to 0$, *and* also conserves entropy/information as we will demonstrate in the next section. However, we have not found a formal proof of the uniqueness of the choice of α_j's.

If one leaves α_1 as a free parameter—does not fix it to the Loop Quantum Gravity value, i.e. $\alpha_1 = -1/2$—, then there is an interesting dividing point in the behavior of the entropy in Eq. (11.10) at $\alpha_1 = -4$. For $\alpha_1 \geq -4$, the entropy in Eq. (11.10) goes to zero, i.e. $S_{BH} = 0$, only at $M = 0$. For $\alpha_1 < -4$, the entropy in Eq. (11.10) goes to zero, i.e. $S_{BH} = 0$, at $M = 0$ and also at some other value $M = M^* > 0$ where M^* satisfies the equation $\frac{4\pi}{\hbar}(M^*)^2 + \pi\alpha_1 \ln\left(1 + \frac{(M^*)^2}{\hbar}\right) = 0$. Thus, depending on the first quantum correction α_1 the black hole mass can vanish if $\alpha_1 \geq -4$, or one can be left with a "remnant" of mass M^* if $\alpha_1 < -4$. It might appear that one could rule out this last possibility since for $M^* > 0$ the black hole would still have a non-zero temperature via Eq. (11.7) and, thus, the black hole should continue to lose mass via evaporation leading to masses $M < M^*$ which would give $S < 0$ for the case when $\alpha_1 < -4$. However, if the Universe has a positive cosmological constant, i.e. space-time is de Sitter, then the Universe will be in a thermal state at the Hawking-Gibbons temperature, i.e. $T_{GH} = \frac{\hbar\sqrt{\Lambda}}{2\pi}$ [16] where $\Lambda > 0$ is the cosmological constant. Thus, if the quantum corrected black hole temperature from Eq. (11.7) becomes equal to

T_{GH} the evaporation process can stop at this finite temperature and still consistently have $S = 0$. This situation would give some interesting and non-trivial connection between the Universal parameter Λ and the final fate of every black hole (in the case when $\alpha_1 < -4$).

Conservation of Energy, Entropy/Information and Solution to the Information Loss Puzzle

We now want to show that the initial (quantum corrected) entropy of the black hole given in Eq. (11.10) can be exactly accounted for by the entropy of the emitted radiation so that entropy/information, i.e. "bits", is conserved. The fact that this happens depends crucially on the specific, logarithmic form of the quantum corrected entropy in Eq. (11.10). This, retrospectively, puts an additional constraint on the α_j's from Eq. (11.9)—other choices of α_j's would not in general lead to both a well behaved S in the $M \to 0$ limit *and* to entropy/information conservation. As we will see, this conservation of information/entropy is connected with the conservation of energy.

To start our analysis, we note that in the picture of Hawking radiation as a tunneling phenomenon the tunneling rate, i.e. Γ, and the change in entropy are related by [2]

$$\Gamma = e^{\Delta S_{BH}}. \tag{11.11}$$

When the black hole of mass M emits a quanta of energy ω energy conservation tells us that the mass of the black hole is reduced to $M - \omega$. Connected with this, the entropy of the black hole will change according to $\Delta S_{BH} = S_{BH}(M-\omega) - S_{BH}(M)$ [9, 10]. Using Eq. (11.10) for the quantum corrected entropy, one obtains for the change in entropy

$$\Delta S_{BH} = -\frac{8\pi}{\hbar}\omega\left(M - \frac{\omega}{2}\right) + \pi\alpha_1 \ln\left[\frac{\hbar + (M-\omega)^2}{\hbar + M^2}\right]. \tag{11.12}$$

Combining Eqs. (11.11) and (11.12), the corrected tunneling rate takes the form

$$\Gamma(M;\omega) = \left(\frac{\hbar + (M-\omega)^2}{\hbar + M^2}\right)^{\pi\alpha_1} \exp\left[-\frac{8\pi}{\hbar}\omega\left(M - \frac{\omega}{2}\right)\right]. \tag{11.13}$$

The term $\exp\left[-\frac{8\pi}{\hbar}\omega\left(M - \frac{\omega}{2}\right)\right]$ represents the result of energy conservation and back reaction on the tunneling rate [9, 10]; the term to the power $\pi\alpha_1$ represents the quantum corrections to all orders in $\hbar$. This result of being able to write the tunneling rate as the product of these two effects, namely back reaction and quantum corrections, depended crucially on the specific form of $S_{BH}(M)$ and ΔS_{BH} from Eqs. (11.10) and (11.12), respectively, which in turn was crucially tied to our specific choice of α_j's in Eq. (11.9). Note that even in the classical limit, where one ignores the quantum corrections by setting $\pi\alpha_1 = 0$, there is a deviation from a thermal spectrum due to the ω^2 term in the exponent in Eq. (11.13).

We now find the connection between the tunneling rate given by Eq. (11.13) and the entropy of the emitted radiation, i.e. S_{rad}. Assuming that the black hole mass is completely radiated away, we have the relationship $M = \omega_1 + \omega_2 + \cdots + \omega_n = \sum_{j=1}^{n} \omega_j$ between the mass of the black hole and the sum of the energies, i.e. ω_j, of the emitted field quanta. The probability for this radiation to occur is given by the following product of Γ's [17–19] which is defined in Eq. (11.13)

$$P_{rad} = \Gamma(M; \omega_1) \times \Gamma(M - \omega_1; \omega_2) \times \cdots \times \Gamma\left(M - \sum_{j=1}^{n-1} \omega_j; \omega_n\right). \tag{11.14}$$

The probability of emission of the individual field quanta of energy ω_j is given by

$$\begin{aligned}
\Gamma(M; \omega_1) &= \left(\frac{\hbar + (M - \omega_1)^2}{\hbar + M^2}\right)^{\pi\alpha_1} \exp\left[-\frac{8\pi}{\hbar}\omega_1\left(M - \frac{\omega_1}{2}\right)\right], \\
\Gamma(M - \omega_1; \omega_2) &= \left(\frac{\hbar + (M - \omega_1 - \omega_2)^2}{\hbar + (M - \omega_1)^2}\right)^{\pi\alpha_1} \\
&\quad \times \exp\left[-\frac{8\pi}{\hbar}\omega_2\left(M - \omega_1 - \frac{\omega_2}{2}\right)\right], \ldots, \\
\Gamma\left(M - \sum_{j=1}^{n-1} \omega_j; \omega_n\right) &= \left(\frac{\hbar + (M - \sum_{j=1}^{n-1} \omega_j - \omega_n)^2}{\hbar + (M - \sum_{j=1}^{n-1} \omega_j)^2}\right)^{\pi\alpha_1} \\
&\quad \times \exp\left[-\frac{8\pi}{\hbar}\omega_n\left(M - \sum_{j=1}^{n-1} \omega_j - \frac{\omega_n}{2}\right)\right] \\
&= \left(\frac{\hbar}{\hbar + (M - \sum_{j=1}^{n-1} \omega_j)^2}\right)^{\pi\alpha_1} \exp(-4\pi\omega_n^2/\hbar).
\end{aligned} \tag{11.15}$$

The Γ's of the form $\Gamma(M - \omega_1 - \omega_2 - \cdots - \omega_{j-1}; \omega_j)$ represent the probability for the emission of a field quantum of energy ω_j with the condition that first the field quanta of energy $\omega_1 + \omega_2 + \cdots + \omega_{j-1}$ have been emitted in sequential order.

Using Eq. (11.15) in Eq. (11.14), we find the total probability for the sequential radiation process described above

$$P_{rad} = \left(\frac{\hbar}{\hbar + M^2}\right)^{\pi\alpha_1} \exp(-4\pi M^2/\hbar). \tag{11.16}$$

The black hole mass could also have been radiated away by a different sequence of field quanta energies, e.g. $\omega_2 + \omega_1 + \cdots + \omega_{n-1} + \omega_n$. Assuming each of these different processes has the same probability, one can count the number of microstates, i.e. Ω, for the above process as $\Omega = 1/P_{rad}$. Then, using the Boltzmann definition of entropy as the natural logarithm of the number of microstates, one gets for the entropy of the emitted radiation

$$S_{rad} = \ln(\Omega) = \ln\left(\frac{1}{P_{rad}}\right) = \frac{4\pi}{\hbar}M^2 + \pi\alpha_1 \ln\left(1 + \frac{M^2}{\hbar}\right). \tag{11.17}$$

This entropy of the emitted radiation is identical to the original entropy of the black hole (see Eq. (11.10)), thus entropy/information/ "bits" are conserved between the initial (black hole plus no radiation) and final (no black hole plus radiated field quanta) states. This implies the same number of microstates between the initial and final states and, thus, unitary evolution. This then provides a possible resolution of the information paradox when the specific conditions are imposed.

The above arguments work even in the case where one ignores the quantum corrections [17–19], i.e. if one lets $\alpha_1 = 0$. While interesting, we are not sure how significant this is since almost certainly quantum corrections will become important as the black mass and entropy go to zero.

In this essay, we have examined the interrelationship of "bits" (information/ entropy) and "its" (physical objects/systems) in the context of black hole information. By requiring that the higher order quantum corrections given in Eq. (11.4) be self-similar in the sense $I_j(r, t) \propto I_0$, and that the associated entropy/information of the black hole as given in Eq. (11.8) be well behaved in the limit when the black hole mass goes to zero, we were able to relate all the higher order quantum corrections as parameterized by the α_j's in terms of the first quantum correction α_1. This proportionality of all α_j's is another level of self-similarity. The final expression for this quantum corrected entropy, namely Eq. (11.10), when combined with energy conservation and the tunneling picture of black hole radiation allow us to show how the original "bits" of black hole information encoded in the horizon were transformed into the "its" of the outgoing correlated Hawking photons, thus providing a potential all orders in $\hbar$ solution to the black hole information loss puzzle.

Finally, as a last comment, it should be stressed that the assumption that the higher order corrections are self-similar in the sense given in Eq. (11.5) (where we take $I_j \propto I_0$) and in Eq. (11.9) (where we take $\alpha_{j+1} \propto \alpha_1$) is not at all what one would expect of the quantum corrections in the canonical approach to quantum gravity where the quantum corrections would in general generate any possible terms consistent with diffeomorphism-invariance. However, this is the problematic aspect of the canonical approach to quantum gravity and, thus, it is worth looking into radical suggestions such as the one proposed here, i.e. that the higher order quantum corrections are greatly simplified by the assumption of self-similarity. This simplification might be seen as an extreme form of the holographic principle of quantum gravity as expounded in [1]. In this monograph, it is pointed out that the entropy of a black hole scales with the area of the horizon while for a normal quantum field theory the entropy will scale as the volume. The conclusion of this observation is that "there are vastly fewer degrees of freedom in quantum gravity than in any QFT" (see chapter 11 of [1]). This assumption of self-similarity of the quantum corrections is in the vein of the holographic principle, since making the assumption of self-similarity means there are vastly fewer types/forms that the quantum corrections can take as compared to canonical quantum gravity.

Acknowledgments There are two works—one on self-similarity [20] and one on the peculiar relationship between long distance/IR scales and short distance/UV scales in quantum gravity [21]—which helped inspire parts of this work.

References

1. L. Susskind, J. Lindesay, *Black Hole, An Introduction To Black Holes, Information And The String Theory Revolution: The Holographic Universe* (World Scientific Publishing Co. Pte. Ltd., Danvers, 2005)
2. M.K. Parikh, F. Wilczek, Hawking radiation as tunneling. Phys. Rev. Lett. **85**, 5042 (2000) [hep-th/9907001]
3. K. Srinivasan, T. Padmanabhan, Particle production and complex path analysis. Phys. Rev. D **60**, 024007 (1999) [gr-qc/9812028]
4. J.D. Bekenstein, Black holes and entropy. Phys. Rev. D **7**, 2333 (1973)
5. H. Von Koch, On a continuous curve without tangents constructible from elementary geometry, in *Classics on Fractals*, ed. by G. Edgar (Addison-Wesley, Reading, 1993), pp. 25–45
6. D. Singleton, E.C. Vagenas, T. Zhu, Insights and possible resolution to the information loss paradox via the tunneling picture. JHEP **1008**, 089 (2010) [Erratum-ibid. **1101**, 021 (2011)] arXiv:1005.3778 [gr-qc]
7. S.W. Hawking, Particle creation by black holes. Commun. Math. Phys. **43**, 199 (1975) [Erratum-ibid. **46**, 206 (1976)]
8. L. Xiang, A note on the black hole remnant. Phys. Lett. B **647**, 207 (2007) [gr-qc/0611028]
9. M.K. Parikh, Energy conservation and Hawking radiation. arXiv:hep-th/0402166
10. M. Arzano, A.J.M. Medved, E.C. Vagenas, Hawking radiation as tunneling through the quantum horizon. JHEP **0509**, 037 (2005) [hep-th/0505266]
11. S.K. Modak, Corrected entropy of BTZ black hole in tunneling approach. Phys. Lett. B **671**, 167 (2009). arXiv:0807.0959 [hep-th]
12. T. Zhu, J.R. Ren, M.F. Li, Corrected entropy of Friedmann-Robertson-Walker universe in tunneling method. JCAP **0908**, 010 (2009). arXiv:0905.1838 [hep-th]
13. T. Zhu, J.R. Ren, M.F. Li, Corrected entropy of high dimensional black holes. arXiv:0906.4194 [hep-th]
14. B. Zwiebach, *A First Course in String Theory* (Cambridge University Press, New York, 2004), p. 221
15. K.A. Meissner, Black hole entropy in loop quantum gravity. Class. Quant. Gravity **21**, 5245 (2004) [gr-qc/0407052]
16. G.W. Gibbons, S.W. Hawking, Cosmological event horizons, thermodynamics, and particle creation. Phys. Rev. D **15**, 2738 (1977)
17. B. Zhang, Q.Y. Cai, L. You, M.S. Zhan, Hidden messenger revealed in hawking radiation: a resolution to the paradox of black hole information loss. Phys. Lett. B **675**, 98 (2009). arXiv:0903.0893 [hep-th]
18. B. Zhang, Q.Y. Cai, M.S. Zhan, L. You, Entropy is conserved in hawking radiation as tunneling: a revisit of the black hole information loss paradox. Ann. Phys. **326**, 350 (2011). arXiv:0906.5033 [hep-th]
19. Y.X. Chen, K.N. Shao, Information loss and entropy conservation in quantum corrected Hawking radiation. Phys. Lett. B **678**, 131 (2009). arXiv:0905.0948 [hep-th]
20. P. Nicolini, B. Niedner, Hausdorff dimension of a particle path in a quantum manifold. Phys. Rev. D **83**, 024017 (2011). arXiv:1009.3267 [gr-qc]
21. E. Spallucci, S. Ansoldi, Regular black holes in UV self-complete quantum gravity. Phys. Lett. B **701**, 471 (2011). arXiv:1101.2760 [hep-th]

Chapter 12
Spacetime Weave—Bit as the Connection Between Its or the Informational Content of Spacetime

Torsten Asselmeyer-Maluga

Abstract In this essay I will discuss the relation between information and spacetime. First I demonstrate that because of diffeomorphism invariance a smooth spacetime contains only a discrete amount of information. Then I directly identify the spacetime as carrier of the Bit, and derive the matter (as It) from the spacetime to get a direct identification of Bit and It. But the picture is stationary up to now. Adding the dynamics is identical to introducing a time coordinate. Next I show that there are two ways to introduce time, the global time leading to quantum objects or the local time leading to a branched structure for the future (tree of the Casson handle). This model would have a tremendous impact on the measurement process. I discuss a model for the measurement of a quantum object with an explicit state reduction (collapse of the wave function) caused by gravitational interaction. Finally I discuss also quantum fluctuations on geometrical grounds. Dedicated to the memory of C.F. von Weizsäcker.

On Bits and Its

In 1990, Wheeler described the concept "it from bit" by the words [21]: IT FROM BIT. OTHERWISE PUT, EVERY 'IT'—EVERY PARTICLE, EVERY FIELD OF FORCE, EVEN THE SPACE-TIME CONTINUUM ITSELF—DERIVES ITS FUNCTION, ITS MEANING, ITS VERY EXISTENCE ENTIRELY—EVEN IF IN SOME CONTEXTS INDIRECTLY—FROM THE APPARATUS-ELICITED ANSWERS TO YES-OR-NO QUESTIONS, BINARY CHOICES, BITS. 'IT FROM BIT' SYMBOLIZES THE IDEA THAT EVERY ITEM OF THE PHYSICAL WORLD HAS AT BOTTOM—A VERY DEEP BOTTOM, IN MOST INSTANCES—AN IMMATERIAL SOURCE AND EXPLANATION; THAT WHICH WE CALL REALITY ARISES IN THE LAST ANALYSIS FROM THE POSING OF YES–NO QUESTIONS AND THE REGISTERING OF EQUIPMENT–EVOKED RESPONSES; IN SHORT, THAT ALL THINGS PHYSICAL ARE INFORMATION–THEORETIC IN ORIGIN AND THAT THIS IS A PARTICIPATORY UNIVERSE. But Wheeler was not the

T. Asselmeyer-Maluga (✉)
German Aerospace Center, Berlin, Germany
e-mail: torsten.asselmeyer-maluga@dlr.de

A. Aguirre et al. (eds.), *It From Bit or Bit From It?*,
The Frontiers Collection, DOI 10.1007/978-3-319-12946-4_12

first. A similar program was carried out by Carl Friedrich von Weizsäcker [19] and his students since the 1950's. Inspired by Heisenberg and Pauli's unified field theory (non-linear SU(2) spinor theory), Weizsäcker considered the simplest bit of quantum information, the ur-alternatives, vectors in the 2-dimensional complex Hilbert space $\mathbb{C}^2$. But the central point in Weizsäckers argumentation is the development of a time-like logic (directly leading to quantum logic) and the relation to probability theory. In particular, he tried to obtain the quantum mechanics by using the ur-alternatives. Here he used 4 approaches to derive the abstract quantum theory (Hilbert space, dynamics). For instance, one approach starts with ur-alternatives and construct a lattice of ur-alternatives leading directly to the Hilbert space. In particular the spacetime is a derived concept in his theory. At this point I disagree with the approach and will discuss a geometric model below. But now let us analyze the two main concepts: the Bit and the It.

So what is information (or the Bit)? Let us look into the standard textbook definition: Information refers to an inherent property concerning the amount of uncertainty for a physical system. First we consider a classical physical system. All information about this system is encoded into the physical state, specified by a distribution function in the multidimensional phase space for all its degrees of freedom. This distribution evolves according to Liouville's theorem, which conserves the phase space volume. At the same it gives rise to the conservation of entropy or information under Hamiltonian dynamics. Now, what is the difference between this and quantum mechanics storing quantum information? For either a pure state is specified by a wave function or a mixed state specified by a density matrix, while Its quantum information content is measured by von Neumann entropy similar to, but structurally equivalent to Shannon entropy. In contrast to classical information, we know that quantum information can neither be cloned nor deleted. In quantum field theory, the information is contained in the state again, a linear functional over an operator algebra. The combination of quantum field theory, general relativity and thermodynamics for a black hole uncovers a problem, the so-called paradox of black hole information loss. The information should be usually conserved in a black hole, where no particle/radiation can be emitted. But Hawking radiation contradicts this conservation of information. Hawking asserted that the emitted radiation from a black hole is thermal and its detailed form is independent of the structure of matter that collapsed to form the black hole. But there is the possibility that the Hawking radiation is entangled with the states in the interior of the black hole, which would solve this paradox.

But what about matter (or the It)? According to the standard model of elementary particle physics, there are quarks and leptons having a rest mass and occupying a non-zero volume (by the Pauli exclusion principle). Furthermore, there are bosons (gluons, W/Z-bosons, photon) mediating the forces between the quarks and leptons. All these constituents can appear in different states. A change of a state is directly caused by an interaction. Above we suggested that the state is the direct expression of information. So, it seems that the It implies the Bit. But conversely, the behavior of the It is controlled by the Bit (caused by interactions). In particular, every outcome of an experiment is a stream of bits and also every dynamics can be seen in this manner. I think Wheeler and Weizsäcker had this picture in mind. But Weizsäcker went further when he introduced time as the steering element in the information stream by using its time-like logic. The discussion of the whole complex behind the slogan 'It from Bit' requires the answer to other questions like what is time? or Is the world digital or continuous? (posted by FQXi in the previous contests).

The Informational Content of the Spacetime

When Einstein developed general relativity (GR), his opinion about the importance of general covariance changed over the years. In 1914, he wrote a joint paper with Grossmann. There, he rejected general covariance by the now famous hole argument. But after a painful year, he again considered general covariance now with the insight that there is no meaning in referring to *the spacetime point A* or *the event A*, without further specifications. Therefore the measurement of a point without a detailed specification of the whole measurement process is meaningless in GR. The reason is simply the diffeomorphism-invariance of GR which has tremendous consequences. Physical observables have to be diffeomorphism-invariant expressions.

The basic object in GR is a smooth 4-manifold M, the spacetime. The (smooth) atlas of M is called the smoothness structure unique up to diffeomorphisms. One would expect that there is only one smooth atlas for any given topological M, all other possibilities can be transformed into each other by a diffeomorphism. But this is not true, see my previous FQXi essay [1]. In fact, there are infinitely many non-equivalent smoothness structures on certain topological Ms with no heuristic to distinguish one above the others as physically relevant. But more importantly, the breakup of the concept 'spacetime point' by using the diffeomorphism invariance is much more important. From the informational point of view, it is the reduction of the continuous information contained in a smooth manifold into a discrete set of relevant subsets. More carefully explained, we divide a smooth manifold into a finite set of simple submanifolds. In topology one calls these submanifolds handles and the division of the manifold its handle decomposition. A k-handle of a n-manifold is the cross product $D^k \times D^{n-k}$ of two disks with $D^k = \{x \in \mathbb{R}^k | \, ||x||^2 \leq 1\}$ having the boundary $\partial D^k = S^{k-1}$ of the $(k-1)$-sphere. Then this k-handle will be glued along $\partial D^k \times D^{n-k}$ to the boundary of a n-disk, i.e. to the $(n-1)$-sphere. To illustrate the power of this concept, I will give an example, the torus $T^2 = S^1 \times S^1$. We start with a 0-handle $D^0 \times D^2$, the disk D^2, and add two 1-handles $D^1 \times D^1$ to the boundary of the 0-handle (see Fig. 12.1). Then we close the manifold by a 2-handle $D^2 \times D^0$ and obtain the torus. In this example we have no freedom in the choice of attaching map for the handle. But adding a 2-handle $D^2 \times D^2$ to build a 4-manifold requires an attaching map $\partial D^2 \times D^2 \to \partial D^4 = S^3$ which can be reduced to $S^1 \to S^3$ (by fixing the second disk D^2). But this map is the definition of a knot! So let us summarize:
A smooth manifold can be decomposed into a diffeomorphism-invariant manner by (at most) countably many handles.
Then the handles can be simply triangulated by using simplices to end up with a piecewise-linear (or PL) structure. The surprising result of Cerf for manifolds

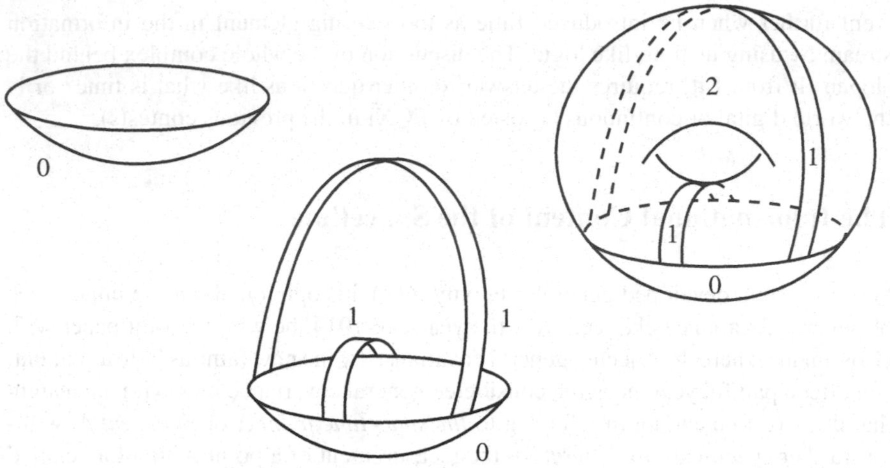

Fig. 12.1 Handle decomposition of the torus

of dimension smaller than 7 was simple: PL-structure (or triangulations) and smoothness structure are the same. This implies that every PL-structure can be smoothed to a smoothness structure and vice versa. Therefore *the discrete approach (via triangulations) and the smooth approach to defining a manifold are the same*! So, our spacetime admits a kind of duality: it contains discrete information in its handle structure but it is a continuous space at the same time. Both approaches are interchangeable.

But an important question remains: Is it possible to obtain this discrete information? Unfortunately, the answer is NO! To understand the core of this answer, I have to introduce an important topological invariant: the fundamental group. Consider all closed curves in a manifold. Two curves are equivalent if the two curves can be continuously deformed into each other (by a so-called homotopy). The equivalence classes of these closed curves forms a group under concatenation, the fundamental group. Beginning with dimension 4, every finitely generated, discrete group can be the fundamental group of a manifold. But then we have the word problem, i.e. for two given fundamental groups we cannot decide whether these groups are isomorphic or not [18]. There is no algorithm for a decision! Or, for two measurements of the fundamental group of the spacetime, we cannot decide whether the two measurements are equivalent. But then we obtain a contradiction to our understanding of an experiment: An replication of the same experiment produces a result but we cannot decide whether it is identical to a previous result.
For two data sets of the spacetime, there is no algorithm to compare the two sets. The result of an experiment is undecidable.
But what is the spacetime in Wheeler's concept? If we do an experiment to measure an observable then we have to choose a coordinate system (a chart in the 4-manifold). Take for example the Stern-Gerlach experiment to measure the spin of an electron. The inhomogeneous magnetic field breaks the isotropy of the space and defines a coordinate system. Then we obtain two streams, electrons with spin $+\frac{1}{2}$ and with

spin $-\frac{1}{2}$ which are space-like separated from each other. Therefore the knowledge of a measurement requires a coordinate system. But spacetime is more. It is the possible set of spacetime points therefore containing all information about coordinates and by the argumentation above in principle also all measurement results. In this spirit, I will state:
The spacetime is the Bit.

From Spacetime to Matter: From Bit to It

In the previous section we discussed the informational content of the spacetime, the Bit. Now I will bridge the gap to the It, the matter. My plan is the derivation of matter from the space or better the geometrization of matter. Unfortunately, this section is the most technical part of the essay. The reader not willing to follow the argumentation can switch to the next section but keeping in mind: *matter and interaction (as gauge theories) can be described as special submanifolds of the space where these submanifolds are determined by the smoothness structure of the spacetime.*

Differential topology is the mathematical theory of smooth manifolds including the (smooth) relations between submanifolds. Let us consider the effect of the change of the smoothness structure (to a non-equivalent one). As an example of this change I consider a compact 4-manifold M (topologically complicated enough, i.e. a K3 surface or more) containing a special torus T_c^2 (so called c-embedded torus). Now cut out a neighborhood $D^2 \times T_c^2$ of this torus (with boundary a 3-torus T^3) and glue in $(S^3 \setminus (D^2 \times K)) \times S^1$ (having also the boundary T^3) where $S^3 \setminus (D^2 \times K)$ denotes the complement of a knot K in the 3-sphere S^3. Then one obtains

$$M_K = \left(M \setminus \left(D^2 \times T_c^2\right)\right) \cup_{T^3} \left((S^3 \setminus (D^2 \times K)) \times S^1\right) \tag{12.1}$$

a new 4-manifold M_K which is homeomorphic to M (Fintushel-Stern knot surgery [16]). If the knot is non-trivial then M_K is not diffeomorphic to M. One calls M_K an exotic 4-manifold, a misleading term. Nothing is really exotic here because all smoothness structures except one (the standard structure) on a 4-manifold are exotic.

What did I change from M to M_K? I simply exchange the torus neighborhood $D^2 \times T_c^2$ by a knot complement $(S^3 \setminus (D^2 \times K)) \times S^1$. Therefore, if I want to understand the smoothness change I have to analyze this knot complement and its effect on the 4-manifold. In [8], we have done this job by starting with the Einstein-Hilbert action at M. Then the change from M to M_K produces some new terms which can be interpreted by using the correspondence between embedded surfaces and spinors. An embedding of a surface in $\mathbb{R}^3$ (up to conformal transformations) is determined by a spinor on this surfaces which fulfills the Dirac equation

$$D\phi = H\phi \tag{12.2}$$

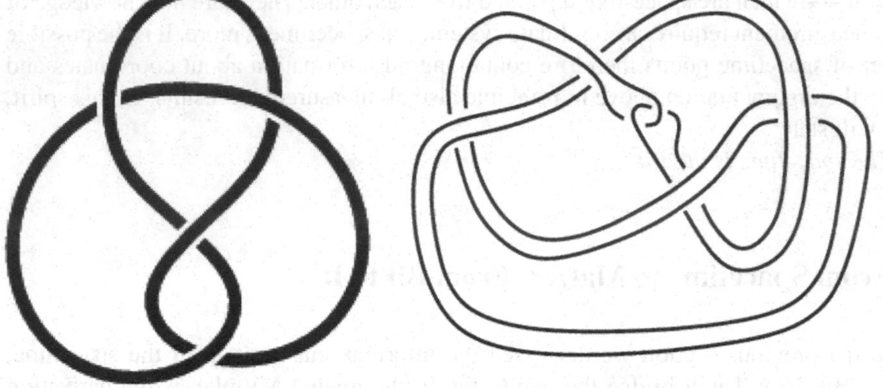

Fig. 12.2 Satellite knot: figure-8 knot (*left*) and the Whitehead double of it (*right*)

with the 2-dimensional Dirac operator D and the mean curvature H of the embedded surface. The Eq. (12.2) looks like an eigenvalue equation to determine the mean curvature. Indeed, one obtains a spectrum of possible geometries for the eigenvalues and any other geometry (or embedding) is a linear combination of eigenvectors. Here, the curvature is quantized without using a quantization of the space (or the spacetime). I do not want to go into the full details but with the help of this theory we were able to derive the Dirac action from the Einstein-Hilbert action. Then the spinor can be directly interpreted as knot complement of the thicken knot $D^2 \times K$ above. But we went a step further and analyze more complex knots, so-called satellite knots (see Fig. 12.2).

Then we obtain a pair of spinors (represented by the two knot complements) which are connected by a torus bundle. A torus bundle can be obtained by taking two copies of $T^2 \times [0, 1]$ and gluing them together. The complexity of this torus bundle depends on the gluing map $T^2 \to T^2$. But there are only three possible gluing maps, so one obtains only three different torus bundles. It was a surprise for us to obtain also the Yang-Mills action in this approach. Then the three types of torus bundles are directly related to the photon, W/Z-bosons and gluons where we automatically obtain the mixing between photon and Z-boson. These results are promising but which knot corresponds to an electron etc.? Here we have only a rough idea to determine the class of knots. A complement of a knot $S^3 \setminus (D^2 \times K)$ is a complex 3-manifold with torus boundary T^2 which can be represented by branched coverings of the 3-sphere. Here I will only make the remark that every 3-manifold can be represented by a 3-fold branched covering of the 3-sphere branched along a knot (a deep theorem of Hilden and Montesinos). Using this result, we are able to determine the knots. At first, the branching set for a knot complement is not a (closed) knot but rather a braid (or knotting strands which are not closed to form a knot). Then, a 3-fold covering induces 3-strand braids as branching set. The braid starts and ends at the boundary so that the interaction can be described by concatenation of braids. This ansatz has many parallels to the Bilson-Thompson model [12] and its extension by Smolin et al. [13]. In an extension of our work [5], the Higgs mechanism was also included.

Time as Regulatory Element: Foliating the Bit to Produce Sequences

In the previous section I unified the Bit and the It in some sense. I derived the It from the Bit but the reverse way is also possible. Now, matter and space have the same root. But I ignored one important element: the order in the Bits. Usually we have sequences of data as the outcome of an experiment. The sequence is an expression of the dynamics and for a given position in the sequence we know the unique precursor and successor. This order structure is denoted as *Time*. But at least with the advent of quantum mechanics we know about the problem of the open future. The outcome of an experiment cannot be known for sure in general. If our spacetime model using exotic smoothness is successful then it should be possible to explain this situation.

The choice of space and time in a spacetime is the determination of a foliation (of codimension one). I remark that Shape dynamics [9, 17] uses also foliations defined in a local way. Standard arguments in GR like causality and Lorentz invariance enforces the choice (up to diffeomorphisms) $\Sigma \times \mathbb{R}$ (see [10, 11]) with a 3-manifold Σ as space. It is also a codimension-one foliation but a global one, i.e. $\Sigma \times \{t\}$ with $t \in \mathbb{R}$ are the (spatial) leafs. An exotic version of $\Sigma \times \mathbb{R}$, denoted by $\Sigma \times_\theta \mathbb{R}$, cannot be (smoothly) foliated in a global manner, see the Fig. 12.3 for an example (the foliation of the torus by infinitely extended planes, the so-called Reeb foliation). It would contradict the exotic smoothness structure: Every 3-manifold Σ has a unique smoothness structure which would imply a unique structure for $\Sigma \times \{t\}$ and therefore for the whole $\Sigma \times \mathbb{R}$. Thus we have to choose a different foliation. Importantly the existence of a codimension-one foliation do not depend on the smoothness structure. In the following I consider the special case of a 3-sphere $\Sigma = S^3$. Then there is no foliation along $\mathbb{R}$ but there is a codimension-one foliation of the 3-sphere S^3 (see [2] for the construction). So, $S^3 \times_\theta \mathbb{R}$ is foliated along S^3 and the leafs are $S_i \times [0, 1]$ with the surfaces $\{S_i\}_{i \in I} \subset S^3$. But otherwise we know that $S^3 \times_\theta \mathbb{R}$ is topologically $S^3 \times \mathbb{R}$. What happens if we enforce a foliation to admit a global time, i.e. with the leafs $S^3 \times \{t\}$? Or equivalently, what happens with the 3-spheres in $S^3 \times_\theta \mathbb{R}$? There is no smoothly embedded S^3 in $S^3 \times_\theta \mathbb{R}$ (otherwise it would have the standard smoothness structure). But there is a wildly embedded S^3! Let $i : K \to M$ be an embedding of K (with $\dim K < \dim M$). One calls the embedding i wild if $i(K)$ is not a finite

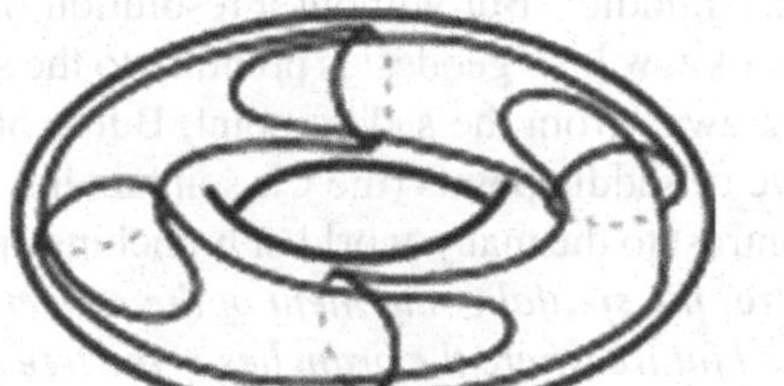

Fig. 12.3 Foliation of the torus (Reeb foliation)

polyhedron (or $i(K)$ is not triangulated by a finite number of simplices). In [3], we considered wildly embedded submanifolds as models of quantum D-branes. The prominent example of a wildly embedded submanifold is Alexanders horned sphere. Wild embedded submanifolds are fractals in a generalized sense. In [6] we argued that this wild embedding is a geometric model for a quantum state. In particular we showed more: the (deformation) quantization of a tame embedding is a wild embedding! If I assume that the spacetime has the right properties for a spacetime picture of quantum gravity then the quantum state must be part of the spacetime or must be geometrically realized in the spacetime. Consider (as in geometrodynamics) a 3-sphere S^3 with metric g. This metric (as state of GR) is modeled on S^3 at every 3-dimensional subspace. If g is a metric of a homogeneous space then one can choose a small coordinate patch. But if g is inhomogeneous then one can use a diffeomorphism to 'concentrate' the inhomogeneity in a chart. Now one combines these infinite charts (I consider only metrics up to diffeomorphisms) into a 3-sphere but without destroying the infinite charts by a diffeomorphism. Wild embeddings are the right structure for this idea. A wild embedding cannot be undone by a diffeomorphism of the embedding space. For the example of Alexanders horned sphere we determine the observable algebra in [6]. It is the hyperfinite factor III_1 von Neumann algebra having the structure of the local algebras in a relativistic QFT with one vacuum vector.

In this model we have one lesson learned: the choice of a global time produces a quantum state (the wildly embedded 3-sphere) but the choice of a local time structure gives a complicated partition of the space. The transition between these two possible foliations is strongly related to the measurement process which will be discussed in the next section. Now I will concentrate on the appearance of time. Above I discussed the foliation problem of exotic $S^3 \times_\theta \mathbb{R}$, i.e. in the terminology of GR this kind of spacetime is not globally hyperbolic. In particular it must contain naked singularities. The structure of these singularities is also known: all singularities are saddle points (see Fig. 12.4 left), i.e. some geodesics meet at the saddle point. This kind of singularity (see Fig. 12.4 middle) has nothing to do with diverging curvatures or metrics. It has a hyperbolic geometry and a finite curvature. The saddle point violates the strong causality in GR but it is what I want. The strong causality in GR is equivalent to a completely deterministic system (like the block universe of Parmenides). If I believe in an open future then I have to introduce the saddle point: some geodesics going to this point whereas some of the geodesics going away from this point (see Fig. 12.4 middle). But without a resolution of the saddle point (see Fig. 12.4 right), I do not know how geodesics pointing to the saddle point are related to the geodesics going away from the saddle point. But exotic smoothness tells us more: the whole weave of saddle points (the Casson handle) in an exotic spacetime forms a tree! So, in contrast to the many-world or branching spacetime interpretation we have another picture: *the spatial component of the spacetime looks like a tree in the time direction called future where the branches of the tree are the possible spatial components.*

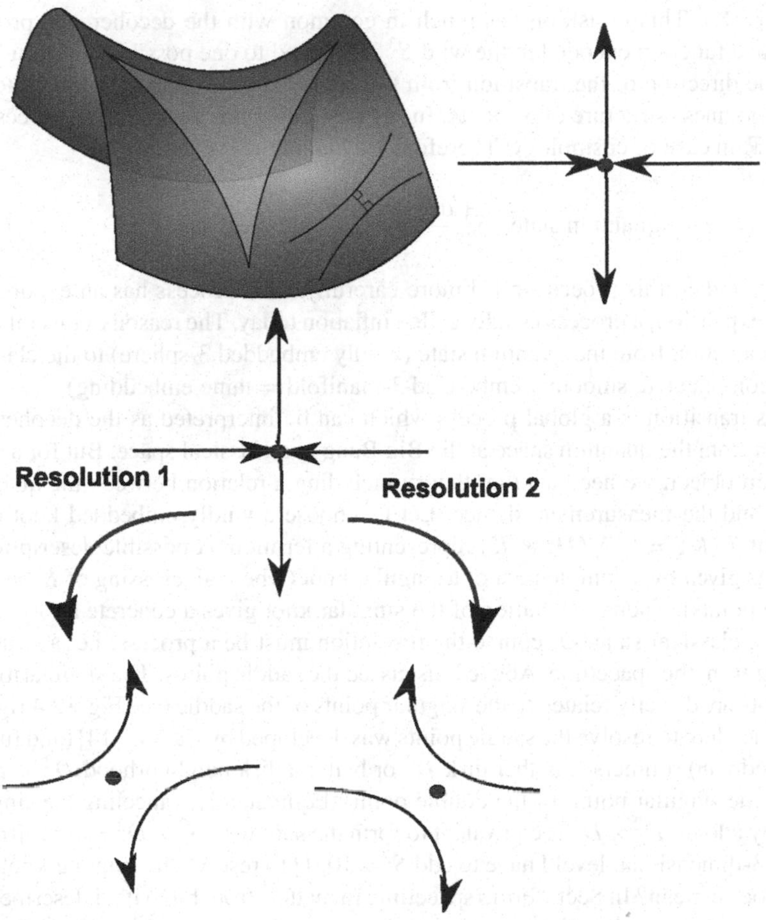

Fig. 12.4 Example of a saddle point (*left*) with the meeting geodesics (*middle*) and the possible resolutions (*right*)

Measurement: Uncovering the Bit

In the description of the exotic $S^3 \times_\theta \mathbb{R}$ by foliations, I introduced the wildly embedded 3-sphere as quantum object (seen as quantization of a tame embedded 3-sphere). A measurement of the quantum object (wildly embedded S^3) should result in a classical space (a tame embedding). The construction of $S^3 \times_\theta \mathbb{R}$ is rather complicated (see [15]). As a main ingredient one needs a homology 3-sphere Σ (i.e. a compact, closed 3-manifold with the homology groups of the 3-sphere) which does not bound a contractable 4-manifold (i.e. a 4-manifold which can be contracted to a point by a smooth homotopy). Interestingly, this homology 3-sphere Σ is smoothly embedded in $S^3 \times_\theta \mathbb{R}$ (as cross section, i.e. $\Sigma \times \{0\} \subset S^3 \times_\theta \mathbb{R}$). But then we obtain a transition from the wild S^3 (quantum object) to a classical space (tame homology

3-sphere Σ). This transition has much in common with the decoherence process. The wave function encoded in the wild S^3 is reduced to one possible state, the tame Σ.[1] The direction of the transition from the wild S^3 to the tame Σ was dictated by the smoothness structure of $S^3 \times_\theta \mathbb{R}$. In [4] we studied this decoherence process for $S^3 \times_\theta \mathbb{R}$ in case of cosmology. Therefore we obtain the transition

$$\text{quantum state} \quad S^3_\theta \overset{\text{decoherence}}{\longrightarrow\longrightarrow\longrightarrow} \text{classical state} \quad \Sigma \tag{12.3}$$

and we studied this process in [7] more carefully. This process has an exponential rate of expansion, a process usually called inflation today. The reason of this inflation is the transition from the quantum state (wildly embedded 3-sphere) to the classical state (complicated, smoothly embedded 3-manifold = tame embedding).

This transition is a global process which can be interpreted as the decoherence process from the quantum space at the Big Bang to a classical space. But for a usual quantum object, we need another theory including a relation between the quantum object and the measurement device. Let us choose a wildly embedded knot complement $\Sigma(K) = S^3 \setminus (D^2 \times K)$ representing a fermion. A possible description of $\Sigma(K)$ is given by a complement of a singular knot (where all crossing of K become double points). Then a resolution of the singular knot gives a concrete knot and we obtain a classical state. Of course the resolution must be a process, i.e. a structure coming from the spacetime. Above I discussed the saddle points. The singularities of the knots are directly related to the singular points of the saddle (see Fig. 12.4 right).[2] The procedure to resolve the saddle points was developed by Casson [14] (and further by Freedman): immerse another disk D^2 or better a disk neighborhood $D^2 \times D^2$ to cancel the singular point (or the double point). Equivalently, canceling the singular point by adding $D^2 \times D^2$ is equivalent to form the sum with $S^2 \times D^2 = S^2 \times [0, 1]^2$. At the 3-dimensional level I have to add $S^2 \times [0, 1]$ to resolve the singular knot. But what does it mean? In Sect. "From spacetime to matter: from bit to it", I described the interaction by torus bundles, complicatedly arranged pieces of $T^2 \times [0, 1]$. By this procedure I obtained the gauge interactions. Now one would expect that gravitation can be also described by a surface bundle. But except torus bundles, there is only one possible bundle, the sphere bundle $S^2 \times [0, 1]$. With these pieces, one can arrange all other possible surface bundles. Now it seems natural to conjecture: *the sphere bundle describes the gravitational interaction*. There are many hints which support this conjecture but no proof. For instance, one can add a sphere bundle to every torus bundle without changing it (universality of gravitation). The gravitational interaction couples to every kind of energy. Therefore one can see gravitation as energy exchange. Let us assume this conjecture then we can *interpret the reason for the reduction of the quantum object (wild knot complement) to the classical state as the*

[1] Here I remark a mathematical fact which is not easy to see: every homology 3-sphere is contained in the wild S^3.

[2] This fact is folklore in Khovanov homology to describe the concordance class of a knot.

gravitational interaction (or the energy exchange) between the measurement device and the quantum object. This idea is not completely new. Penrose was the first who notice it but without proof. Of course the whole process is only a proposal but it follows directly from our geometric model.

Smearing the Bit: Quantum Fluctuations

A quantum fluctuation is the temporary appearance of energetic particles out of empty space as controlled by the famous Heisenberg uncertainty principle. In quantum field theory, the interior lines of the Feynman diagrams are the expression for these fluctuations (also called virtual particles). The appearance of quantum fluctuations is a stochastic (or better uncontrolled) element in the description. How does this concept fit into our geometrical picture? Above, I spoke about foliations, i.e. the division of spacetime into space (the leaf) and time at least locally. Foliations of exotic 4-manifolds (like $S^3 \times_\theta \mathbb{R}$) have a fundamental property: nearby leafs for a time t can be far away from each other for later times. Mathematically, the generating dynamics of the foliation has a chaotic behavior: a small perturbation of the initial condition (the choice of a special leaf) results in a large deviation. An example of such a foliation can be seen in Fig. 12.5, a foliation (or lamination) of a disk by hyperbolic geodesics. This relation to chaotic dynamics is not accidental, because a wild embedding is a fractal and the self-similarity is a sign for this behavior.

Fig. 12.5 (Geodesic) lamination of the disk

Conclusion

I have presented a certain number of ideas and results:

1. Because of diffeomorphism invariance, spacetime seen as smooth 4-manifold contains only a discrete amount of information. Spacetime itself is the Bit.
2. There is a freedom in the definition of the spacetime coming from the choice of the smoothness structure. This idea can be used to identify some submanifolds (related to the smoothness structure) with the matter (fermions and bosons). The It is equal to the Bit.
3. For example consider the foliation of an exotic spacetime like $S^3 \times_\theta \mathbb{R}$ can be very complicated. But the structure of the foliation uncovers the structure of the time. Time is a regulatory element. The past is determined but the future is open.
4. For the usual foliation $S^3 \times \{t\}$ with $t \in \mathbb{R}$ of $S^3 \times_\theta \mathbb{R}$ the 3-sphere must be a wildly embedded submanifold (represented by an infinite polyhedron).
5. A quantum state can be defined on the spacetime as wild embedding. A wild embedding can be seen as a quantization of a tame embedding.
6. This identification between quantum state and wild embedding has a strong impact to understand the measurement process. So, I discussed the possibility that gravitation enforces the state reduction after a measurement.
7. Quantum fluctuations have also a geometrical background by using the (hyperbolic) foliation.

We will end up this essay with two quotations. At first Wheelers words [20] about Time and its meaning: TIME, AMONG ALL CONCEPTS IN THE WORLD OF PHYSICS, PUTS UP THE GREATEST RESISTANCE TO BEING DETHRONED FROM IDEAL CONTINUUM TO THE WORLD OF THE DISCRETE, OF INFORMATION, OF BITS. … OF ALL OBSTACLES TO A THOROUGHLY PENETRATING ACCOUNT OF EXISTENCE, NONE LOOMS UP MORE DISARMINGLY THAN 'TIME.' EXPLAIN TIME? NOT WITHOUT EXPLAINING EXISTENCE. EXPLAIN EXISTENCE? NOT WITHOUT EXPLAINING TIME. TO UNCOVER THE DEEP AND HIDDEN CONNECTION BETWEEN TIME AND EXISTENCE … IS A TASK FOR THE FUTURE.

And Secondly, a quote from the famous Argentine writer Jorge Luis Borges who wrote in his short story "The Garden of Forking Path": THE GARDEN OF FORKING PATHS IS AN INCOMPLETE, BUT NOT FALSE, IMAGE OF THE UNIVERSE AS TS'UI PN CONCEIVED IT. IN CONTRAST TO NEWTON AND SCHOPENHAUER, YOUR ANCESTOR DID NOT BELIEVE IN A UNIFORM, ABSOLUTE TIME. HE BELIEVED IN AN INFINITE SERIES OF TIMES, IN A GROWING, DIZZYING NET OF DIVERGENT, CONVERGENT AND PARALLEL TIMES. THIS NETWORK OF TIMES WHICH APPROACHED ONE ANOTHER, FORKED, BROKE OFF, OR WERE UNAWARE OF ONE ANOTHER FOR CENTURIES, EMBRACES ALL POSSIBILITIES OF TIME. WE DO NOT EXIST IN THE MAJORITY OF THESE TIMES; IN SOME YOU EXIST, AND NOT I; IN OTHERS I, AND NOT YOU; IN OTHERS, BOTH OF US. IN THE PRESENT ONE, WHICH A FAVORABLE FATE HAS GRANTED ME, YOU HAVE ARRIVED AT MY

HOUSE; IN ANOTHER, WHILE CROSSING THE GARDEN, YOU FOUND ME DEAD; IN STILL ANOTHER, I UTTER THESE SAME WORDS, BUT I AM A MISTAKE, A GHOST.

Both quote shows the necessity to study the phenomenon of time. My own thoughts are more in the direction of Borges where the tree of the Casson handle is an expression of the decision tree after a quantum measurement.

Before concluding, I must add that the views expressed are only partly original. I have partially drawn from the ideas of Carl H. Brans, Jerzy Król and Helge Rosé. I was also strongly influenced by the work of C.F. von Weizsäcker.

References

1. T. Asselmeyer-Maluga, A chicken-and-egg problem: which came first, the quantum state or spacetime? See (2012). http://fqxi.org/community/forum/topic/1424 (Fourth Prize of the FQXi Essay contest "Questioning the Foundations" (see http://fqxi.org/community/essay/winners/2012.1))
2. T. Asselmeyer-Maluga, J. Król, Abelian Gerbes, generalized geometries and exotic R^4. submitted to J. Math. Phys. (2009). arXiv:0904.1276
3. T. Asselmeyer-Maluga, J. Król, Topological quantum d-branes and wild embeddings from exotic smooth R^4. Int. J. Mod. Phys. A **26**: 3421–3437 (2011). arXiv:1105.1557
4. T. Asselmeyer-Maluga, J. Król, Decoherence in quantum cosmology and the cosmological constant. Mod. Phys. Lett. A **28**: 1350158 (2013). doi:10.1142/S0217732313501587, arXiv:1309.7206
5. T. Asselmeyer-Maluga, J. Król, Higgs potential and confinement in Yang-Mills theory on exotic R^4 (2013). arXiv:1303.1632
6. T. Asselmeyer-Maluga, J. Król, Quantum geometry and wild embeddings as quantum states. Int. J. Geom. Methods in Mod. Phys. **10**(10) (2013) will be published in November 2013. arXiv:1211.3012
7. T. Asselmeyer-Maluga, J. Król, Inflation and topological phase transition driven by exotic smoothness. Adv. HEP, **2014**: 867460 (article ID) (2014). http://dx.doi.org/10.1155/2014/867460, arXiv:1401.4815
8. T. Asselmeyer-Maluga, H. Rosé, On the geometrization of matter by exotic smoothness. Gen. Rel. Grav. **44**, 2825–2856 (2012). doi:10.1007/s10714-012-1419-3, arXiv:1006.2230
9. J. Barbour, B. Bertotti, Mach's principle and the structure of dynamical theories. Proc. R. Soc. Lond. A **382**: 295–306 (1982)
10. A.N. Bernal, M. Sánchez, On smooth Cauchy hypersurfaces and Geroch's splitting theorem. Commun. Math. Phys. **243**: 461–470 (2003). arXiv:gr-qc/0306108
11. A.N. Bernal, M. Saánchez, Globally hyperbolic spacetimes can be defined as "causal" instead of "strongly causal". Class. Quant. Gravity, 24:745–750 (2007). arXiv:gr-qc/0611138
12. S.O. Bilson-Thompson, A topological model of composite preons (2005) arXiv:hep-ph/0503213v2
13. S.O. Bilson-Thompson, F. Markopoulou, L. Smolin, Quantum gravity and the standard model. Class. Quant. Gravity **24**, 3975–3994 (2007). arXiv:hep-th/0603022v2
14. A. Casson, Three lectures on new infinite constructions in 4-dimensional manifolds, vol 62. (Birkhäuser, progress in mathematics edition, 1986. Notes by Lucian Guillou, first published 1973)
15. M.H. Freedman, A fake $S^3 \times R$. Ann. Math. **110**, 177–201 (1979)
16. R. Fintushel, R. Stern, Knots, links, and 4-manifolds. Inv. Math. **134**: 363–400 (1998). (dg-ga/9612014)

17. H. Gomes, T. Koslowski, The link between general relativity and shape dynamics. Class. Quant. Gravity **29**, 075009 (2012). arXiv:1101.5974
18. A. Markov, Unsolvability of certain problems in topology. Dokl. Akad. Nauk SSSR **123**, 978–980 (1958)
19. C.F. Von Weizsäcker, The unity of nature. Farrar, Straus, and Giroux, New York, 1980. Most books like: Aufbau der Physik (1985) or Zeit und Wissen (1992) in German
20. J.A. Wheeler, Hermann Weyl and the unity of knowledge. Am. Sci. **74**, 366–375 (1986)
21. J.A. Wheeler, Information, physics, quantum: The search for links. Complexity, Entropy, and the Physics of Information. In: W. Zurek, ed, Addison-Wesley (1990)

Chapter 13
Now Broadcasting in Planck Definition

Craig Hogan

Abstract If reality has finite information content, space has finite fidelity. The quantum wave function that encodes spatial relationships may be limited to information that can be transmitted in a "Planck broadcast", with a bandwidth given by the inverse of the Planck time, about 2×10^{43} bit/s. Such a quantum system can resemble classical space-time on large scales, but locality emerges only gradually and imperfectly. Massive bodies are never perfectly at rest, but very slightly and slowly fluctuate in transverse position, with a spectrum of variation given by the Planck time. This distinctive new kind of noise associated with quantum geometry would not have been noticed up to now, but may be detectable in a new kind of experiment.

At the turn of the last century, Max Planck derived from first principles a universal formula for the spectrum of radiation emitted by opaque matter. Planck's radiation law solved a long-standing experimental mystery unexplained by classical physics, and agreed exactly with measurements. It flowed from a simple, powerful and radically new idea: that everything that happens in nature occurs in discrete minimum packages of action, or quanta. Planck's breakthrough started the quantum revolution in physics that defined much of twentieth century science and technology.

A few years after Planck's triumph, Albert Einstein introduced his theory of relativity. While Planck's theory addressed the nature of matter, Einstein's addressed the nature of space and time. It also solved long standing mysteries, and flowed from a simple idea: that the laws of physics should not depend on how one moves. Einstein extended his theory with another powerful idea—that local physics is the same in any freely falling frame—to reveal that space and time form an active dynamical geometry, whose curvature creates the force of gravity. General Relativity was revolutionary, but it is entirely classical: Einstein's space-time is not a quantum system.

These two great theories of twentieth century physics have never been fully reconciled, because their core ideas are incompatible. Relativity is based on the notion of locality, a concept not respected by quantum physics; indeed, experiments with

C. Hogan (✉)
University of Chicago and Fermilab, Chicago, USA
e-mail: cjhogan@fnal.gov

A. Aguirre et al. (eds.), *It From Bit or Bit From It?*,
The Frontiers Collection, DOI 10.1007/978-3-319-12946-4_13

quantum systems prove that states in reality are not localized in space. The central role of measurement in quantum physics flies against the relativistic notion that reality is independent of an observer. Perhaps most fundamentally, relativity violates quantum precepts by assigning tangible reality to unobservable things, such as events and paths in space-time.

This clash of ideas led to agonized epistemological debates in the early part of the century, most famously between Bohr and Einstein. But most of physics has moved on. For all practical purposes so far, it works just fine to assume a continuous classical space and put quantum matter into it. That is what the well-tested Standard Model of physics does. It is a quantum theory, but only of matter, not of space-time.

Classical continuous space, as usually assumed, maps onto real numbers: it has an infinite information density. Quantum theory suggests instead that the information content of the world is fundamentally limited. It is natural to suppose that all spatial relationships are just another sort of observable relationship, to be derived from the quantum theory of some system. Let us adopt a working hypothesis different from the usual one: *information in spatial position is limited by the broadcast capacity of exchanged information at a bandwidth given by a fundamental scale*. It is possible to work out some experimental consequences of this hypothesis even in the absence of a full theory, because the fidelity of space is limited by its information capacity.

We have some clues to the amount of information involved. Planck's formula came with a new constant of nature, a fundamental unit for the quantum of action that we now call Planck's constant, $\hbar$. By combining his constant with Newton's constant of gravity G and the speed of light c, Planck obtained new "natural" units of length, time and mass. In Einstein's theory, G controls the dynamics of space and time. Planck's units therefore set the natural scale for the quantum mechanics of space-time itself, where information about location becomes fundamentally discrete.

Because gravity is weak, the Plank scale is very small; for example, the Planck time is $t_P = \sqrt{\hbar G/c^5} = 5.4 \times 10^{-44}$ s. And because that scale is so small, its quanta are very fine grained, and so far undetectable. No experiment shows an identifiable quantum behavior of space and time. That is why for the last century, physicists have been able to treat space and time like a definite, continuous, classical medium.

The lack of an experiment means that we have no guide to interpret mathematical ideas about blending quantum mechanics and space-time. Physicists were forced into the strange world of quantum mechanics by experimental measurements, such as radiation spectra from black bodies and gases. As Rabi said, "Physics is an experimental science." So, let us ask a very practical question: *How can we build an experiment that directly reveals the discrete character of space-time information at the Planck scale?* The answer depends on the character of that information—the encoding of quantum geometry in macroscopic position states.

Consider how quantum systems work. In pre-quantum physics, all properties of a system, such as positions and velocities of particles, have definite values, and change with time according to definite rules. In quantum physics, as the state of a system evolves, relationships among its properties evolve according to definite rules—but in general, individual properties do not have definite values, even in principle.

Instead, the entire system is described by a wave function of possibilities. Reality is that multitude of possibilities, a set of relationships. In general, definite, observable outcomes are impossible to predict.

Any combined system is literally more than the sum of its parts; a composite system contains information that cannot be separated into information about one subsystem or another. Information in a combined system generally resides in the correlation between its parts, a property known as "quantum entanglement".

The quantum challenges to conventional notions of what is real were highlighted in a famous 1935 paper by Einstein, Podolsky, and Rosen. Schrödinger responded by introducing the idea of entanglement, as well as the provocative thought-experiment with the uranium, the flask of hydrocyanic acid, and the unlucky cat. As he noted, entanglement comes with nonlocality: "Maximal knowledge of a total system does not necessarily include total knowledge of all its parts, not even when these are fully separated from each other and at the moment are not influencing each other at all" [1]. Einstein referred to such behavior as "spooky action at a distance".

Although these ideas have created controversy over the years, it is an experimental fact that information in the real world is not localized in space and time. A measurement in one place is correlated with, and affects the state of a system everywhere [2, 3]. There are even real-world experiments that show examples of such quantum entanglement between particles that never co-existed at the same time [4]. Although experiments that demonstrate such effects are quite subtle to mount, entanglement, and the nonlocality that goes with it, are woven into the fabric of reality.

Einstein and others realized that quantum nonlocality is a big problem for relativity [5]. It seems to directly contradict the foundational notion of space and time, that everything happens at a definite time and place. Even the most advanced theory of space-time, General Relativity, is based on a metric that specifies the intervals between events. Quantum mechanics implies that intervals between events, or indeed any property of events themselves, can never be exactly measured, even in principle. And some things that happen in the world—the quantum correlations created by interactions that we interpret as the collapse of a wave function—are entirely delocalized in space and time.

Physics has advanced by working around the apparent paradox of delocalization. Often, it is not important to know exactly where something happens, since many important properties of matter do not depend on location in any particular place. For example, in quantum field theory, the quantized system is a mode of a field wave extended in space and time—a delocalized state. This approximation leads to extraordinarily successful predictions for all experiments on collisions of elementary particles at energies far below the Planck mass, such as those at Fermilab and CERN.

Quantum delocalization inspires a view of the world made not so much of material as of information. This idea may be extended to space and time as well as matter. Some properties of space and time that seem fundamental, including localization, may actually emerge only as a macroscopic approximation, from the flow of information in a quantum system.

Even within the entirely classical framework of relativity and gravitation, theory has provided some hints about the possibility of such emergence, and about the significance of the Planck scale—how quantum mechanics blends with the physics of space and time. The purest states of space-time, black holes, obey thermodynamic properties: for example, in a system of black holes, the total area of event horizons always increases, like entropy. More generally, the equations of general relativity can be derived from a purely statistical theory, by requiring that entropy is always proportional to horizon area [6]. Similarly, Newton's laws of motion and gravity can be derived from a statistical theory based on entropy and coarse-graining, where information lives on surfaces and is associated with position of bodies [7]. In these derivations, the dynamics of space-time, the equivalence principal, and concepts of inertia and momentum for massive bodies, all arise as emergent properties.

These results are based on essentially classical statistical arguments, but they refer to quantum information. The detailed quantum character of the underlying quantum degrees of freedom associated with position in space is not known, but we can guess some of their properties. The precision and universality of light propagation, even across cosmic distances [8], suggests that causal structure arises from a fundamental symmetry, even if locality is only approximate. Gravitational thermodynamics also suggest an exact number for the distribution and amount of information in the system: the information fits on two-dimensional sheets, and the total information density is the area of a sheet in Planck units. From these arguments, we do not know how this holographic information is encoded, but we know how much of it there is, and something about how it maps onto space.

Taken together, these theoretical ideas hint that quantum mechanics limits the amount of information in space-time. Presumably, such a limit must place some kind of limit on the fidelity of space-time itself: not all classically described locations are physically different from each other.

Video buffs know that higher bandwidth gives you a better picture. Suppose that the bandwidth of information transmission is limited by the Planck frequency: $\omega_P \equiv t_P^{-1} \equiv \sqrt{c^5/\hbar G} = 1.85 \times 10^{43}$ Hz. If this is the best that the cosmic Internet Service Provider can give us, we do not get a perfect picture. There is only a finite amount of information in the positional relationships of material bodies. Perhaps a very careful experiment, that looks at position closely in the right way, might be able to see a bit of blurring, a lack of sharpness and clarity, like a little extra noise in the image or clipping in the cosmic sound track. The properties of the clipping may even reveal something about the compression or encoding algorithm.

An effect like that would of course be interesting to physicists, who are essentially hobbyists of nature. We used to say that physics is about discovering laws of nature, but these days we could just as well say that it is all about figuring out how the system of the universe works—how its instructions are encoded, and what operating system it runs on. An experiment would provide some useful clues.

Imagine then that the real world is the ultimate 4-dimensional video display. How good is it?

At first you might guess that the Planck bandwidth limit would simply create a system with Planck size pixels everywhere; that is, a frame refresh rate given by the Planck time t_P, and pixel (or voxel) size given by the Planck length, $ct_P \equiv \sqrt{\hbar G/c^3} = 1.6 \times 10^{-35}$ m, in each of the three space dimensions.

To achieve such a fine grained picture, we need a Planck bandwidth channel for every pixel– a Planck density of information in four dimensions, or a Planck bandwidth in every three dimensional Planck volume. This value is the amount of information in the standard model of quantum fields, if we include all frequencies up to the Planck scale.

But this guess does not agree with holographic emergence. Our radically different hypothesis is that space and time are created from information propagating with Planck bandwidth. In such a "Planck broadcast", space is not assumed to exist *a priori*, but is a set of relationships that emerges from Planck-limited information processing. Instead of a world densely packed with Planck size cells, as in field theory, perhaps positions in space and time only contain the amount of information that can be carried on a Planck frequency carrier wave. In that case, a large spatial volume has a much smaller density of information.

Imagine that someone sends broadcast video at the Planck frequency. That is only enough information to refresh one pixel every Planck time. For a larger screen, the refresh rate and resolution get worse. How much worse?

If the broadcast encodes all of real space, it needs to encode all directions. Suppose that the video screen is a sphere of with a radius L about our broadcast point, and has pixels of size Δx. We encode the information on the screen to refresh more slowly, with a refresh interval given by the time it takes light to get to the screen and back, $\tau = 2L/c$—the slowest acceptable rate for encoding a position at this distance. Then the minimum pixel size is given by setting the total number of pixels $4\pi L^2 \Delta x^{-2}$ per time $2L/c$ equal to the Planck information rate t_P^{-1}, so the pixel size is $\Delta x = \sqrt{2\pi L c t_P}$—very small, but still much larger than the Planck length.

Of course, nature is not really pixelated in little squares, but the same answer for the blurring scale emerges from a more realistic physical model based on waves. Positions encoded by wave functions that have a cutoff or bandwidth limit convey only a limited amount of transverse spatial information from one place to another [9, 10].

Imagine a wave that passes through a pair of narrow slits. The wave creates an interference pattern on a screen at distance L that depends on (i.e., encodes) the transverse separation of the two slits. However, there is a resolution limit: if the two slits have transverse separation much smaller than

$$\Delta x_\perp \approx \sqrt{L c t_P}, \tag{13.1}$$

the interference pattern of radiation at frequency ω_P is not distinguishable from that of a single slit. The resolution limit from this point of view is a diffraction limit in wave mechanics, but it is really an information bound: the waves simply do not have enough information to resolve smaller transverse distances than that. Notice that the distance to the screen—and causal structure—can be defined with much higher precision $\approx c t_P$, by counting wave fronts. The transverse resolution, the slit separation (Eq. 13.1), gets much poorer at large L.

The corresponding angular uncertainty,

$$\Delta\theta = \Delta x_\perp / L \approx \sqrt{ct_P/L}, \tag{13.2}$$

gets smaller on large scales. Thus, angles get *sharper* at larger separation, so the notion of direction emerges more clearly on larger scales. The total amount of angular information grows, but only linearly with L, more slowly than it would for a display with Planck size pixels.

Overall there are about L/ct_P degrees of freedom corresponding to radial separation, and L/ct_P corresponding to angle. For each Planck time in duration or radial separation, there are L/ct_P angular degrees of freedom. The total amount of information is the number of directions times the duration, so it grows holographically, like $(L/ct_P)^2$. The density of information is constant on surfaces, but in 3D space it thins out with time and distance as it spreads.

This holographic scaling is just what is needed for the statistics of emergent gravity to work. If we invoke that idea to set the scale of information density, the prediction for transverse mean square position uncertainty becomes very precise [11]:

$$\langle \hat{x}_\perp^2 \rangle = Lct_P/\sqrt{4\pi} = (2.135 \times 10^{-18}\text{m})^2 (L/1\text{m}), \tag{13.3}$$

with no free parameters. We don't know the character of the actual quantum theory that controls geometry, but this estimate of the transverse blurring scale is relatively robust, because it is just determined by the amount of information.

Apparently, if space-time is a quantum system with limited information—a Planck broadcast—there should be a new kind of quantum fuzziness of positions, not just for small particles, but for everything, even for large masses. The blurring is larger for larger L: the position resolution gets worse at larger distances. In a laboratory size system, it is much larger than the Planck length—about an attometer in scale, a billionth of a billionth of a meter.

There is vastly less information in this macroscopic quantum system than in standard theory—that is, a system of quantum fields in classical space-time with a Planck cut off—but there is enough angular information to agree with the apparent sharpness and classical behavior of space, as measured in experiments to date. If things could be measured at separations on the Planck scale, the angular uncertainty would be huge; directions are not even really well defined, and it is essentially a 2D holographic system. On the scale explored by particle colliders, about 10^{16} times larger than Planck length, things are already very close to classical; angular blurring is too small to detect with particle experiments of limited precision, and in any case the particle masses are small so standard quantum effects overwhelm the geometrical ones.

Indeed, the new Planck blurring is *always* negligible compared to standard quantum uncertainty (which does depend on mass) for systems much smaller than about a Planck mass, $m_P = \sqrt{\hbar G/c} = 2.176 \times 10^{-8}$ kg [11]. In measurements of small numbers of particles, the geometrical effects are not detectable. Unlike standard quantum effects, the Planck information limit is only important for *large* masses.

At first, it also seems strange that the resolution depends on a macroscopic separation. Intuition suggests that the state of affairs of matter and energy should not depend on how far away it is; after all, how can it "know" where we, the Planck broadcaster or observer, are? According to Einstein, the laws of physics ought to be independent of the location and motion of an observer.

A related worry is that an attometer scale uncertainty, while small, is really not all that small by the standards of particle physics. That scale is now routinely resolved by particle colliders, like the Tevatron and the LHC. Yet there is no sign in experiments of a new kind of fuzziness in space-time. Indeed, if we set L comparable to the size the universe, we find that Δx is actually on a scale you can see with your own eyes, of the order of 0.1 mm, the width of a hair. Space certainly doesn't display any lack of sharpness on that scale when you look around.

These worries may be resolved by invoking entanglement. Space-time is the ultimate, universal entangled system. Locality itself can emerge, via entanglement, as an approximate behavior on large scales.

Information is not localized in space, but resides in non localized correlations. The density of information can depend on scale, and can be smaller for larger systems. The effective fidelity of space-time can change depending on where something is relative to an observer. A measurement confined to a small volume does not know or care about a transverse geometrical displacement relative to some distant place, so the uncertainty is not observable in local measurements.

In quantum mechanics, measurements make projections—in Copenhagen language, they "collapse" the wave function. Until they are made, there is uncertainty given by the width of the wave function—in our case, the scale-dependent blurring. In an emergent space-time, every world-line defines a particular projection of the wave function associated with the structure of nested light cones (or "nested causal diamonds") around it.

Thus, the quantum geometrical position information is entangled for bodies whose world-lines are close together. If you measure the transverse position state of one massive body, you will find almost the same projection of the geometrical state for any body nearby. That does not mean that any two bodies are in the same position; it only means that their quantum deviation from the classical position is almost the same, relative to any arbitrary far away point. The local relationships of the bodies in space are changed very little from standard quantum mechanics.

A classical space-time is the limiting case of a fully coherent system. The approach to the classical limit however reveals slight departures from classical behavior that are not present in standard theory. Nonlocal projections of the quantum state in different directions are slightly different, even on large scales.

As the system unfolds in time, the uncertainty leads to random variations—a new kind of noise in position measurements to distant bodies in different directions. The positions of nearby bodies change together, carried along with the geometry, into the same new definite state. Local measurements are not affected by this collective change of position—a new kind of "movement without motion".

This interpretation of the angular uncertainty opens up a way to build an experiment that probes Planck scale physics. The Planck broadcast model of quantum geometry predicts that positions fluctuate, with a power spectrum of angular variations given approximately by the Planck time—that is, in an average over duration τ, the mean square variation is

$$\langle \Delta\theta^2 \rangle_\tau \approx t_P/\tau. \tag{13.4}$$

In an experiment of size L, the variations accumulate up to durations $\tau \approx L/c$, ultimately leading to variance in position given by the overall uncertainty, Eq. (13.3). This prediction can be tested by making very sensitive measurements of transverse positions of massive bodies. The measurement process must make a nonlocal comparison of position in different directions.

An experiment designed to detect or rule out fluctuations with these properties, called the Fermilab Holometer, is currently being developed [12]. It uses a technique based on laser interferometers like those used to measure gravitational waves. The intensity of light emerging from a Michelson interferometer allows a precise and coherent measurement of the positions of mirrors over an extended region of space, in this case, 40 m in two directions. The precision of such devices is extraordinary; they can detect variations in mean position differences on the order of attometers, limited primarily by the quantum character of the laser light. In the Holometer, correlations are measured between the signals of two adjacent, aligned interferometers. The correlations are sensitive to tiny, random in-common motions that change very quickly, on timescales comparable to a light-crossing time, less than a microsecond. (On longer timescales, entanglement-driven locality reduces the variation). The effective speed of the motionless movement is tiny—comparable to continental drift, only centimeters per year.

Because of quantum entanglement, the holographic noise created by the Planck broadcast information limit creates tiny, rapid fluctuations in signals from the two adjacent interferometers that are coherent with each other, even if there is no connection between the devices apart from proximity. Arguments like those outlined here, based on information in holographic emergent space, can be used to make an exact prediction for the expected cross-correlated noise spectrum, even without knowing details of the fundamental theory [11, 13].

Whether or not new Planck scale holographic noise is detected, the Holometer is interesting as an exploratory experiment, because it tests the fidelity and coherence of nonlocal spatial relationships with Planck precision for the first time. The outcome will either reveal a signature of new Planck scale physics, or experimentally prove a coherence of macroscopic space greater than what is possible with a Planck broadcast. We don't know what we will find.

References

1. E. Schrödinger, Naturwissenschaften **23**, 807 (1935) (Trans. J. Trimmer)
2. R. Horodecki, P. Horodecki, M. Horodecki, K. Horodecki, Rev. Mod. Phys. **81**, 865 (2009)
3. X. Ma, S. Zotter, J. Kofler, R. Ursin, T. Jennewein, C. Brukner, A. Zeilinger, Nat. Phys. **8**, 480–485 (2012)
4. E. Megidish, A. Halevy, T. Shacham, T. Dvir, L. Dovrat, H.S. Eisenberg, Phys. Rev. Lett. **110**, 210403 (2013)
5. E.P. Wigner, Rev. Mod. Phys. **29**, 255 (1957)
6. T. Jacobson, Phys. Rev. Lett. **75**, 1260 (1995)
7. E.P. Verlinde, JHEP **1104**, 029 (2011)
8. A.A. Abdo et al., Nature **462**, 331–334 (2009)
9. C.J. Hogan, Phys. Rev. D **77**, 104031 (2008)
10. C.J. Hogan, Phys. Rev. D **78**, 087501 (2008)
11. C.J. Hogan, A model of macroscopic quantum geometry, arXiv:1204.5948 [gr-qc]
12. http://holometer.fnal.gov
13. C.J. Hogan, Phys. Rev. D **85**, 064007 (2012)

Chapter 14
Is Spacetime Countable?

Sean Gryb and Marc Ngui

Not everything that counts can be counted and not everything that can be counted counts.

–Albert Einstein

Abstract Is there a number for every bit of spacetime, or is spacetime smooth like the real line? The ultimate fate of a quantum theory of gravity might depend on it. The troublesome infinities of quantum gravity can be cured by assuming that spacetime comes in countable, discrete pieces which one could simulate on a computer. But, perhaps there is another way? In this essay, we propose a picture where scale is meaningless so that there can be no minimum length and, hence, no fundamental discreteness. In this picture, Einstein's Special Relativity, suitably modified to accommodate an expanding Universe, can be reinterpreted as a theory where only the instantaneous shapes of configurations count.

Counting What Counts

This essay is about what things we can count, and what we can't. Practicalities won't concern us. It may be very difficult, for example, to count the number of grains of sand on a beach, or the number of molecules in our body, or even the number of quantum states of our brain; but, in principle, these things can be done. We're also not even concerned with whether the number of things to be counted is finite or not. Even if we have to go on counting *forever*, if each element of a set can be given a number, then we are happy to call that set *countable*.

S. Gryb (✉)
Institute for Mathematics, Astrophysics and Particle Physics,
Radboud University Nijmegen, Nijmegen, The Netherlands
e-mail: sean.gryb@gmail.com

M. Ngui
Bumblenut Pictures, Toronto, Canada

A. Aguirre et al. (eds.), *It From Bit or Bit From It?*,
The Frontiers Collection, DOI 10.1007/978-3-319-12946-4_14

You would think that most everything would be countable; but, unexpected things are provably not. For example, it's impossible to count the number of provable theorems in a mathematical theory. That's a theorem. Another thing that can't be counted is the number of degrees of freedom in a field—like the fields that constitute all our most basic theories of physics. This fact causes headaches (which we will discuss soon) but, perhaps surprisingly, it doesn't prevent us from very accurately describing Nature. Indeed, it would appear that our basic understanding of physics and mathematics relies exclusively on things that cannot be counted. But are we right to think this, or is there something incomplete about our current understanding of physics? Are field theories really fundamental or do we need a new framework for making sense of our world? Another way to phrase this is ask whether the degrees of freedom that make up Nature are *discrete*, like the bits in a computer, and can be counted with the natural numbers, or whether they are *continuous* like the elements of the real line. We will present a scenario where physics must be continuous because, at a fundamental level, it is scale invariant. If scale doesn't exist, there can be no minimum length (because this minimum length would provide a preferred scale in the theory) and, therefore, no discreteness and no way to fundamentally capture the physics of our world on a standard computer. We will show how observers in Einstein's special theory of Relativity can be reinterpreted as observers in a scale-invariant space. This relationship appears to be intimately linked with a new formulation of gravity called *Shape Dynamics* [1], which we will come back to at the end. For the moment, we can look for a clue for how to make pragmatic progress on these issues by considering the nature of the gravitational force.

I already mentioned the infinity of degrees of freedom in a field theory and the headaches they cause for physicists and mathematicians. These headaches are most commonly dealt with using a framework called *renormalization*. Renormalization works on the principle that a theory behaves in a different way depending on how accurate your measuring procedure is. Some theories, like the field theories that describe the forces important for atoms, nuclei, and nuclear constituents, behave in an increasingly simple way when the measuring procedure becomes more and more accurate. For these theories, a finite number of measurements need to be performed for the parameters of the theory (like the masses of the particles or the relative strength of the force) to be determined. These theories are called *renormalizable* and are deemed acceptable field theories because, once the parameters have been obtained, the result of any measurement can be predicted once the accuracy of the measurement is specified. Unfortunately, the simplest quantum theory of gravity is not a theory of this type.

There is a simple way to understand why this is true. In General Relativity, energy warps spacetime and there is no limit to the amount of warping that is possible. This means that regions of very dense energy can collapse under their own weight to form regions of infinitely curved spacetime. Near these singular points, nothing can escape and these now familiar regions are called *black holes*. In Quantum Theory, if you make a measurement of an object's position, its momentum inevitably becomes more uncertain. Thus, by accurately measuring the position of an object, you create a high probability that this object will have a correspondingly large energy.

Now put General Relativity and Quantum Theory together. Doing this, you can reach a point where you can measure the position of a particle so accurately that its average energy becomes large enough to produce a black hole that is bigger than the region in which you are trying to detect that very particle. Gravity produces a black hole that ruins your quantum measurement. The length scale at which this happens is called the *Plank scale* and understanding what happens to physics at these scales is one of the great mysteries of modern physics.

There are several strategies for attacking this problem. The two most common are listed below:

- *Introduce new physics*: It may be that General Relativity is not the correct theory of gravity and that a new theory, with nicer quantum properties, takes over at length scales that we have not yet been able to probe experimentally.
- *Abandon the continuum*: If spacetime is fundamentally discrete—that is, if there are only a countable number of degrees of freedom in the theory—and this discreteness presents itself before the Plank scale is reached, then the theory is cured because the problematic region has been eliminated.

The two most studied modern approaches to quantum gravity, String Theory and Loop Quantum Gravity, make use of these strategies; the former using the first strategy and the latter the second.

Although the research programs following these two strategies have made impressive progress, important open questions still remain. For example, approaches that try to introduce new physics inevitably run into the problem that General Relativity is a very robust theory, so that it is difficult to modify it without ruining its basic structures. One is then presented with many ambiguities for how to do this and these ambiguities are not easy to resolve without conflicting with known experiments. On the other hand, the "fundamental discreteness" scenario seems to suffer from a rather immediate drawback. Since no mathematical framework with a countable number of elements can ever be proven to be finished (because of the theorem I mentioned earlier), there is no way to prove that your "fundamental" theory is ever truly fundamental. It is impossible to know for certain whether some other theory is not underlying the true behaviour of the system. It is not known whether frameworks based on the continuum are subject to a similar restriction, but perhaps they can be proven to be superior in this regard.

Given these and other open questions in the standard approaches, it is perhaps justified to consider other strategies. One such strategy embraces the continuum and requires that physics should be fundamentally scale invariant at its most basic level. In approaches that follow this strategy, there can be no notion of discreteness because a minimum length scale would be quite obviously in conflict with the requirement that scale is meaningless.

The Case for Scale Invariance

In the theory of renormalization, when continuous fields are considered, one of the most common ways for a particular theory to be renormalizable is for it to be scale invariant at its most fundamental level—that is, when infinitely accurate (or high energy) measurements are considered.[1] This makes sense because a scale invariant limit of this kind means that the theory itself eventually stops changing at a certain point as you keep making your measurements more precise. This provides you with a kind of anchor that allows you to determine what the theory should look like when you start making coarser and coarser measurements. If the anchor isn't fixed, the theory could drift anywhere. However, there is an even more basic reason for wanting fundamental scale invariance in your theory: only dimensionless quantities have objective meaning. A "meter" doesn't have any meaning on its own unless it is compared against the length of another object. Thus, it is only in the scale-invariant description of the theory that its parameters can be sensibly given an objective, dimensionless value.

For these reasons, there exists a number of approaches that aim to describe gravity in a scale-invariant way, either exactly or in some high energy limit of the theory. This task would seem difficult because scale seems to be an important part of our description of modern physics. Neither the Standard Model (our current framework for understanding the sub-atomic physics) nor General Relativity are manifestly scale invariant. Nevertheless, there are several approaches that aim to achieve a scale-invariant description of Nature. The most direct of these are approaches that aim to describe gravity and matter directly in terms of locally scale-invariant physics. Such approaches were originally pursued by Weyl [2] and recently by authors such as 't Hooft [3] or Mannheim [4]. Other approaches aim to recover an approximate notion of scale invariance in the high energy limit of the theory.[2] These include the asymptotic safety program [5], which aims to make sense of quantum General Relativity as a quantum field theory. Unfortunately, despite many years of effort, none of these approaches have provided a completely adequate picture of scale-invariant gravity.

What all of these approaches have in common, is that they are considering a *spacetime* notion of scale invariance. In this work, we will instead be concerned with a slightly different notion: that of *spatial* scale invariance. We will now present a framework which suggests that *spatial* scale invariance may actually be hidden in the framework of Special Relativity, which forms the starting point for General Relativity. Indeed, we will give an argument showing how the concepts of spacetime and spatial scale invariance can be interchanged using well-known mathematical transformations. This interchangeability seems to be intimately connected with a new formulation of General Relativity, called *Shape Dynamics*, where local scale invariance is manifest.

---

[1] The technical requirement is that the theory have an *ultra-violet fixed point*.

[2] I am referring to the search for a UV fixed point (which has scale invariance in form of vanishing beta-functions) in the theory space of General Relativity.

Scale Invariance in an Expanding Universe

The Expanding Universe

That the spacetime description of Special Relativity can be traded for a scale-invariant description of separated space and time can be made possible by considering two ingredients:

a. That space is closed and has the shape of a 3 dimensional sphere. This means that an observer can head in the same direction and (eventually) come back to their original location.
b. That space is expanding and that this expansion if fueled by a very small, positive parameter called the *cosmological constant*, which we will discuss briefly.

The first ingredient is an assumption of simplicity. If true, then a holistic picture of the world is, at least in principle, possible. The second ingredient is taken from observation, and must be part of our description of reality. Note that what is actually observed as evidence of expansion is what is called the *red shift*, which means that the frequency of light appears to get stretched out over time. We measure this by comparing *ratios* of lengths, which gives a dimensionless (i.e., scale-invariant) number. Thus, the concept of expansion can be well-defined in a way that does make reference to any absolute scales. It is this second, observationally motivated ingredient that will be key to our argument. This is a new input in that it was unknown to Einstein and others during the development of Relativity.

We will now add these two postulates to Einstein's original postulates of Special Relativity:

1. The Laws of physics should take the same form for any *inertial observer*.[3]
2. The speed light is defined to be a finite constant, c, for all inertial observers.

The first postulate is an assumption of simplicity, while the second is proposed for compatibility with observations. Just as in Special Relativity, we will assume the existence of idealized rods and clocks which can be perfectly synchronized. An inertial observer can label any event that can occur using a coordinate indicated by these idealized clocks and rods using some prescribed procedure (which won't interest us here). If space is not expanding, it is then an easy exercise to show, using Einstein's postulates, that different inertial observers will register a different set of coordinates for the same event and that these coordinates are related through a set of transformations called the *Poincaré transformations*. We then say that the physical events are *Poincaré invariant*.

Adding the two postulates (a) and (b) changes things considerably. We would like to imagine how the Universe would be expanding under the influence of the observed cosmological constant if the influences of all other forms of matter and energy could be ignored. This assumption is, in fact, not valid but we will still be able to apply our

[3] An *inertial observer* is one that is not moving under the influence of an external force.

model to the real world if we take this as a *local principle* for constructing a more realistic theory. That means that we should think of our "expanding Universe" below as what an observer would see in a sufficiently small region near the location of the observer.

We know what such an expanding Universe should look like because we understand how the Universe would expand in the presence of a cosmological constant only. What happens is as follows. Space has the shape of a 3 dimensional sphere, and at any time this sphere has some radius. To be precise, let's call that radius, r. In time, this radius changes its size. For an inertial observer who is stationary in some coordinate system, r changes according to a very specific rule: the square of r is equal to the square of the time interval read on the observer's clock (times c) plus the square of the cosmological horizon that we will call ℓ (and which is related to the cosmological *constant* by taking its inverse square):

$$r^2 = (ct)^2 + \ell^2. \tag{14.1}$$

The resulting spacetime is curved and has the shape of a hyperboloid. We will call this a *de Sitter (dS)* hyperboloid after the first person to study its properties. We can visualize a curved dS hyperboloid by drawing it inside of a flat spacetime in one extra dimension. This is completely analogous to how one can visualize a 2 dimensional curved sphere by drawing it inside a 3 dimensional flat space, even though the third dimension is not accessible to observers confined to live on the surface of the sphere. In Fig. 14.1a, we show what the dS hyperboloid looks like. Time flows upwards. Since, for visualization purposes, we have added an extra spatial dimension, these are labeled by (x, y, z, w) and we have collapsed the (x, y, z)-direction and have only shown the x-direction. These spatial coordinates should not be confused with the three *real* spatial coordinates (θ, ϕ, ψ) which are periodic and can be chosen to represent angles on a 3 dimensional sphere. Just as observers on a sphere cannot move into the regions interior and exterior of the sphere, an observer in dS space cannot move into the regions interior and exterior to the hyperboloid. As can be seen from Fig. 14.1a, in the infinite past, space has a larger and larger size. This shrinks down to the minimum value of ℓ before beginning an expansion phase that continues into the infinite future. An important feature of the dS hyperboloid are its inertial observers. They follow the "straight lines" on the hyperboloid, which are represented by hyperbola that extend from the distant past to the distant future. A typical observer of this kind is illustrated in light blue in Fig. 14.1b.

A *stationary* observer, is an observer whose (θ, ϕ, ψ)-coordinates on the sphere do not change in time, t. We know, however, from the fact that space is expanding, that the fictitious (x, y, z, w)-coordinates must change in time in order for the relation (14.1) to be maintained. By convention, we can pick the w-direction to be the only direction that is changing. Thus, for the stationary observer, $x = y = z = 0$. Events which occur in the dS hyperboloid are distinguished by points to which the stationary observer will attribute a particular set of coordinates. Other inertial observers are "straight lines" on the hyperboloid in the sense that they follow the extreme paths on the hyperboloid.

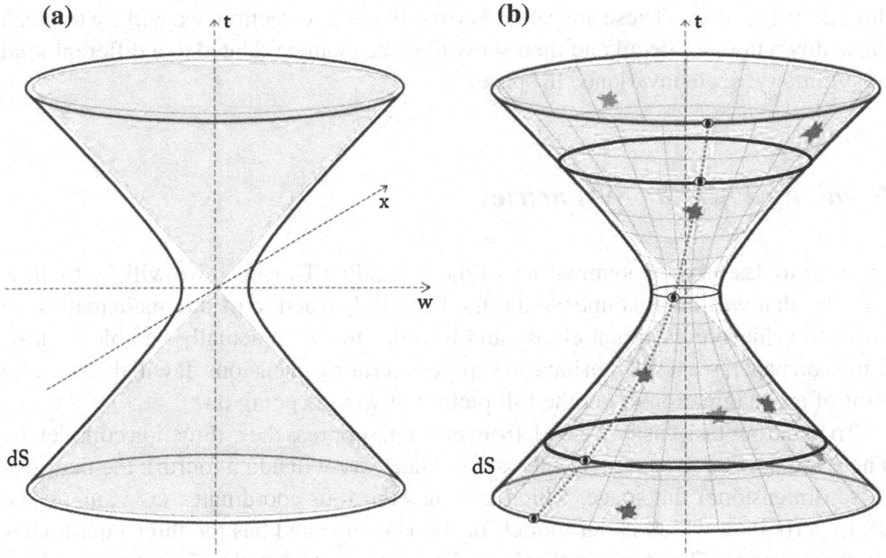

Fig. 14.1 The de Sitter hyperbola. Observers are depicted as eyes and events as bursts of light. **a** dS hyperboloid. **b** A typical inertial observer

Now we are in a position to understand how Einstein's theory of Special Relativity can be extended to an expanding Universe. We want to be able to relate the coordinates attributed to events in the stationary observer's reference frame to the coordinates attributed to the *same* events in the reference frame of some other inertial observer. We are thus looking for the set of transformations that generalize the Poincaré transformations in a dS Universe. In order for Einstein's first principle to hold, these transformations must preserve the shape of the dS hyperboloid so that the form of the Laws of physics are unchanged. In other words, we are looking for the set of symmetry transformations of the dS hyperboloid. Mathematically, this corresponds to the set of transformations that preserve the relation (14.1) because this is the defining relation of the hyperboloid. Since the hyperboloid itself is unchanged, the distance between two points remains the same so that inertial observers will continue to be inertial observers after the transformations.

It is now a rather straightforward mathematical exercise to identify the set of transformations that preserve the form of (14.1). Let's refer to them as the group of dS symmetries. In total, there are 10 of them (the same number as the original Poincaré transformations in flat space (x, y, z, t)-spacetime). There are six attributed to the symmetries of the 3 dimensional sphere: three of which are rotations and three are translations of the (θ, ϕ, ψ) coordinates. These are analogous to the translations and rotations in the familiar flat (x, y, z)-space that we learn about in high-school. The other four are associated with time. One is just a time translation. The remaining three are what happens when you change your velocity in either of the three possible directions (θ, ϕ, ψ). These are called *boosts*. In the next section, we will try to break these down in more detail and then show how they can be related to a different kind of symmetry: scale invariance in space.

Breaking down the Symmetries

In order to describe the symmetries of the expanding Universe, we will try to draw pictures that will help us understand the beautiful structure of the mathematics. In order to achieve conceptual clarity and in order for us to actually be able to draw things on paper, we will often have to suppress certain dimensions. It will then require a bit of an imagination to get the full picture of what is going on.

To make our task easier we will, from now on, suppress the z-direction completely. There are now three spaces of interest (and later, we will add a fourth): the fictitious extra dimensional flat space, which now has the four coordinates (x, y, w, t); the dS hyperboloid, which is our model for the Universe and has the three coordinates (θ, ϕ, t); and the 2 dimensional sphere with coordinates (θ, ϕ). Here are how these spaces are related. As already discussed, the dS hyperboloid is the surface in the fictitious flat space whose points obey the relation (14.1). The 2-spheres can be obtained by drawing surfaces of constant t. These are the planes drawn in Fig. 14.2a and they intersect the dS hyperboloid in what look like circles. However, these are not actually circles because we've suppressed one spatial dimension. Remember that the r-direction is actually the xy-plane. This means that, what looks like circles, are actually 2 dimensional spheres, which, unfortunately, we can't draw in our limited number of dimensions. Figure 14.2b shows how we can imagine the 2-spheres which intersect the planes of constant t. These are "snapshots" of the Universe and they are clearly changing in size over time.

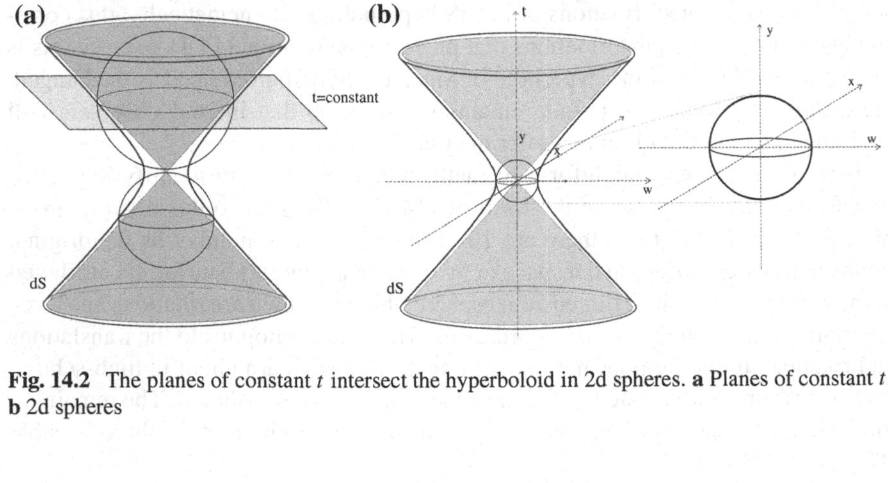

Fig. 14.2 The planes of constant t intersect the hyperboloid in 2d spheres. **a** Planes of constant t. **b** 2d spheres

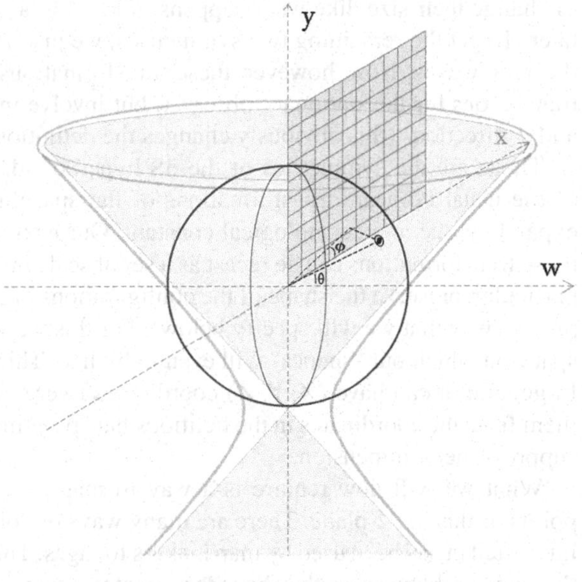

Fig. 14.3 The polar angles θ and ϕ on the 2-sphere

Lets examine these 2d spheres. Remember our convention where we called the w-direction the direction where our stationary observer is moving. This suggests that we choose a spherical coordinate system where the w-direction points up. It is then customary to choose two angles, θ and ϕ, that parameterize the 2d sphere. The ϕ variable goes from 0 to 2π and represents the angle in the xy-plane. The θ variable goes from 0 to π and represents the angle to the w-axis (as is shown in Fig. 14.3). We can see that we now have consistency with our postulate (a) since, as advertised, our (θ, ϕ) variables are periodic

$$\theta = \theta + 2\pi \qquad \phi = \phi + 2\pi. \tag{14.2}$$

We can now visualize the symmetries of the dS hyperboloid. The simplest ones are the ones associated with the symmetries of the 2-sphere. Since we are suppressing one dimension, we now have three: one rotation and two translations. The rotation is the familiar rotation of the sphere that keeps the North pole fixed; i.e., it is a rotation about the w-axis. The two translations are the two different ways to move the North pole. These involve independent or simultaneous shifts of the θ and ϕ coordinates.

The symmetries associated with time are a bit harder to visualize, especially because the size of the spheres is changing in time. It is perhaps simplest to visualize by how they act on the hyperboloid in the flat, higher dimensional space. In this space, they look like a kind of time-space rotation of the hyperboloid around one of the spatial axes. Take, for example, the tw-"rotation" of the hyperboloid about the xy-axis. This is easiest to visualize from the point of view of the stationary observer at $x = y = 0$. This observer follows the hyperbola given by the intersection of the xy-plane with the dS hyperboloid. Thus, the tw-"rotation" just pushes the observer up in time along its trajectory. The only effect of this transformation is for the 2-spheres to change their size, like what happens in Fig. 14.2a. This will be an important fact later. To get the remaining two symmetries, we just have to replace the w-axis with the x or y axes. Now, however, these transformations do not look like simple time translations for the stationary observer, but involve increasing the velocity in the θ and ϕ directions (this obviously changes the definition of our stationary observers).

Those are the symmetries of the dS hyperboloid. They are the generalizations of the usual Poincaré transformations of flat spacetime to the case of a Universe expanding due to a cosmological constant. Our next, and final, task is to show that these transformations can be recast as a set of scale invariant transformations that, in particular, preserve the shape of the configurations of observers in the Universe (in a precise way that we will specify below). For this we will need to introduce one last space on which our "shapes" will eventually live. This is just a flat Euclidean plane. In general, it can have (X, Y, Z) coordinates (were we used capitals to distinguish them from the coordinates in the fictitious flat spacetime), but, for simplicity, we will suppress the Z-dimension.

What we will now require is a way to map the points on the 2-sphere to the points of this flat 2-plane. There are many ways of doing this but the one we will be interested in has been used by map makers for ages. This is because map makers have the same problem as us: they have to project locations on the round Earth to points on a flat map. The technique used by navigators is called the *stereographic projection* and the same property that makes it useful for navigation will also be useful for us: it preservers angles! The way to perform the stereographic projection is to imagine a light bulb siting on the South pole of the sphere (i.e., the point $x = y = 0$ and $w = -\sqrt{\ell^2 + (ct)^2}$). Then, imagine placing the plane so that it is tangent to the sphere at the North pole (see Fig. 14.4). The shadow cast on the plane by a point on the sphere is its stereographic projection. It's clear that any point on the sphere (except the South pole) will have a stereographic projection onto the plane.

We now return to the key property of the stereographic projection: any angle formed by the intersection of two lines on the sphere will be preserved by the projection. Consider a particular $t =$ const 2-sphere in a model Universe where a stationary observer is making observations in the presence of two other inertial observers. At this instant, one can draw imaginary lines between each of the particles forming a kind of triangle. Under the stereographic projection, the angles of this triangle are preserved. Because an observer can only makes measurements locally, the angles they measure are the only objective way for them to determine the "shape" of this three particle configuration. We can then say that the shape is preserved under the projection.

By "scale invariance", we mean that the theory doesn't depend on the size of the configurations of the system. Instead, only the shapes, as defined above, should be important. This kind of scale invariance is also called *conformal invariance*, and the transformations that preserve angles (or the shapes of the instantaneous configurations of the system) are called *conformal transformations*. Again, it is a relatively straightforward mathematical exercise to determine what these conformal transformations are. In 2 dimensions, they can be written mathematically using the complex variable

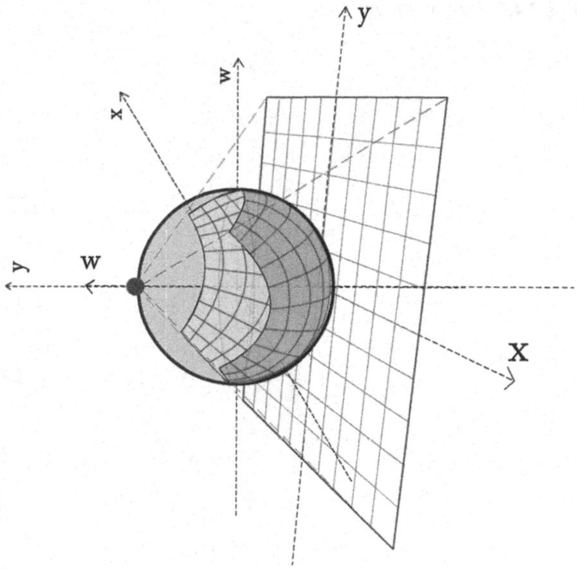

Fig. 14.4 The stereographic projection of a grid onto the plane

$$\zeta = X + iY, \tag{14.3}$$

where $i = \sqrt{-1}$. In terms of this variable, the conformal transformations are also called *Möbius transformations*. The mathematical definition of the Möbius transformations is given in Appendix A. Here we will describe them physically.

There are six different kinds Möbius transformations. The simplest involve no changes of scale at all. There are three of these. The first, is a rotation in the XY-plane and the other two are translations in the X and Y directions. Clearly, these won't change the shape of the system. The other three involve changes in the global scale. The first of these are dilatations where only the global scale of the system is changed. The last two are a bit harder to visualize. They are called *special conformal transformations* and can be visualized most easily by imagining how they can be stereographically projected onto the plane from the sphere. A special conformal transformation is a combination of translating the position of the sphere over the plane and performing an *inversion*, which involves rotating the position of the North pole as shown in Fig. 14.5. In fact, all the Möbius transformations can be represented in a simple way using the stereographic projection.

One may have noticed a similarity between the symmetry group of the dS Universe and the Möbius transformations. In both cases, there are six transformations and, in both cases, they can be represented by how they act on surfaces of constant t. Indeed—and this is the key observation necessary for our analysis—the Möbius transformations can be shown to be equivalent to the dS symmetries when the expansion of space starts to become large! This happens during the very early and very late times of our Universe. Thus, the behaviour of inertial observers during these epochs

Fig. 14.5 An inversion

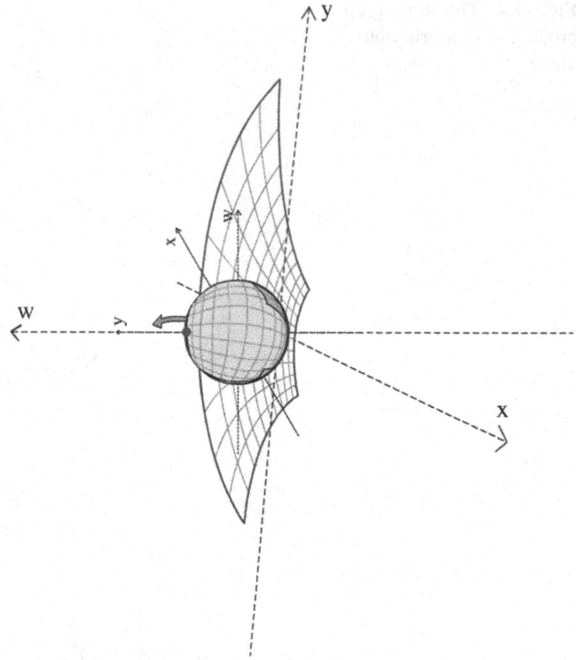

can be used to link the spacetime description of events to a scale-invariant one. Using this correspondence, we can map a system of inertial observers in the dS Universe to conformally invariant point particles (with a fixed time parametrization) in flat space. A concrete prescription for doing this is given in Appendix B. Many more details about the model presented here can be found in the technical paper [6] which feeds heavily off of the mathematical structures discussed in [7].

This entire construction relies upon being able to represent the dS symmetries, in the distant past and future, as conformal transformations on the plane via the stereographic projection. We will now illustrate how this can be done for the two simplest of these transformations. For a brief description of this, consult Appendix B or, for the full computation, see [6]. The simplest transformations to relate are the rotations around the w-axis in dS space. These are quite obviously equivalent to rotations in the XY-plane after stereographic projection. The second transformation, which is slightly more non-trivial, is the time translation for the stationary observer. As we pointed out, this transformation corresponds to simple time translations for the stationary observer. Since the size of the 2-spheres changes in time, these correspond to dilatations on the XY-plane. Figure 14.6 shows how this happens. The remaining Möbius transformations are harder to visualize because they are combinations of translations and boosts. Furthermore, these can only be shown to be equivalent in the distant future and distant past (this leads to the *holographic* nature of the construction presented in the Appendix B). We encourage the reader to try to work these out for themselves.

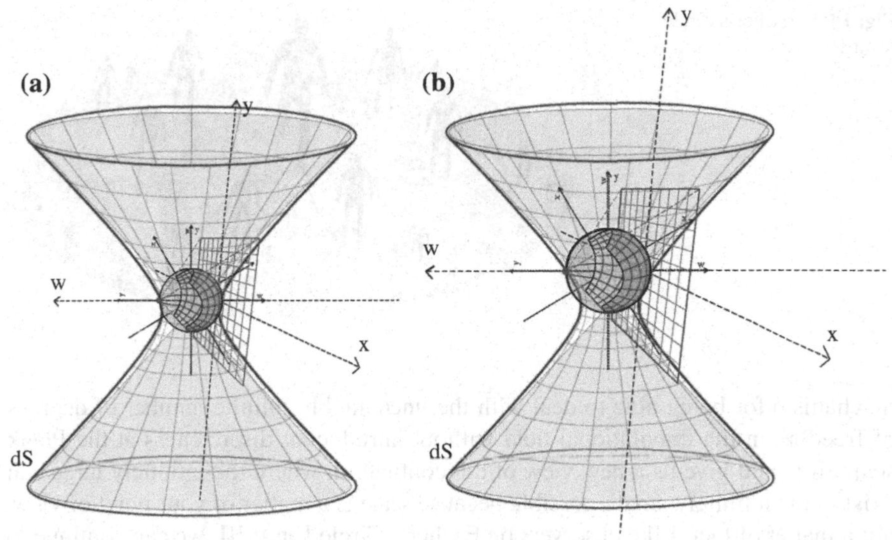

Fig. 14.6 In time, the projection of the grid grows in size. **a** The project at time t. **b** The project at a later time

Conclusion

We have shown how it is possible to reinterpret the trajectories of inertial observers in dS space in terms of the trajectories of particles where only the instantaneous shape of the spatial configurations count. The (asymptotic) symmetries of dS space can be transformed to the conformal symmetries on the plane via a stereographic projection. The picture we describe here could help clarify how to understand the meaning of inertial observers in Shape Dynamics, which is a new framework for gravity where *local* scale is traded for the time part of spacetime symmetries.

In this essay, it has been the *global* aspects of scale invariance that have concerned us. However, in a full theory of gravity, it is *local* Poincaré, and the corresponding *local* scale, transformations that are relevant. For that more complicated case, one can build up a curved spacetime by gluing together local patches of flat spacetime using the techniques of *Cartan geometry* [8]. Could it be possible to do something similar in Shape Dynamics; i.e., glue together local patches of *conformally* flat spaces that could then be related to General Relativity through the correspondence outlined here? A first step towards doing this was performed in $2+1$ dimensions [9], but many additional difficulties arise in the more physical case of $3+1$ dimensions. Some further investigations towards this end have been explored by Derek Wise in [7]. It is clear that further insights are needed to sort out these intriguing possibilities, but the relationships described here could be a first step towards achieving such insights.

But what does all of this suggest? Perhaps it suggests that there is a way to think of quantum gravity in fully scale-invariant terms. If true, this would provide a new

Fig. 14.7 Escher's circle limit III

mechanism for being able to deal with the uncountably infinite number of degrees of freedom in the gravitational field without introducing discreteness at the Plank scale. It would give us a new view of the continuum, where the infinitely large can exist in the infinitely small, possible because scale is a matter of your point of view, not a matter of fact. Like observers on Escher's Circle Limit III, we can continue to peer into the infinite complexities of our world, ever pondering the mysteries that lie beyond (Fig 14.7).

Möbius Transformations and the Lorentz Transformations

The Möbius transformations are defined as:

$$\zeta \to \frac{a\zeta + b}{c\zeta + d}, \tag{14.4}$$

where a, b, c, d are complex numbers obeying $ac - bd \neq 0$. This group is well-known to be isomorphic to the projective special linear group PSL(2,), which, in turn, is isomorphic to the orthochronous Lorentz group $SO^+(3, 1)$. It is this property that we exploit in Appendix B. For more info on the Möbius transformations and for visualizations which inspired our diagrams on stereographic projection, see [10].

De Sitter Inertial Observers to Scale Invariant Particles

For a much more detailed account of the material presented here, see the technical paper [6].

We are inspired by the Shape Dynamics formulation of gravity, as presented in [1], where equivalence with GR is manifest in Constant Mean Curvature (CMC) slicings of solutions to the Einstein equations. For $dS^{d,1}$ spacetime, the CMC slices are constant t hypersurfaces in the ambient $^{d+1,1}$ and have $\mathcal{S}^d$ topology. To see this, we can use a convenient choice of coordinates for the embedding:

$$t = \ell \sinh \varphi \qquad x^I = \ell \cosh \varphi \, \tilde{x}^I , \tag{14.5}$$

where $I = 1, \ldots, (d+1)$ and $\tilde{x}^I \tilde{x}^J \delta_{IJ} = \tilde{x}^2 = 1$. Using these coordinates, the induced metric is

$$ds^2 = -\ell^2 d\varphi^2 + \ell^2 \cosh^2 \varphi \, d\Omega^2 , \tag{14.6}$$

where $d\Omega^2$ is the line element on the unit d-sphere. Since the spatial metric is conformal to the metric on the unit sphere (which is homogeneous), it is clear that this slicing must be CMC.

We now consider a useful set of coordinates:

$$x^{\pm} = x^0 \pm x^{d+1} \qquad X^i = \frac{x^i}{x^0 - x^{d+1}} , \tag{14.7}$$

where $i = 1, \ldots, d$. The $x^{\pm}$ are just light-cone coordinates in the ambient space. We can single out one of these, namely x^-, as a convenient time variable and write the other $x^+ = \frac{x^2 - \ell^2}{x^0 - x^{d+1}} = \frac{1}{x^-}\left(\frac{X^2}{(x^-)^2} - \ell^2\right)$ using the definition of de Sitter spacetime. The X^i's are a convenient choice of spatial coordinates because, as can be shown with a straightforward calculation, in the limit as $t \to \pm\infty$ (i.e., the conformal boundary of spacetime), they are just giving the stereographic projection of coordinates on the constant-t hypersurfaces onto a Euclidean plane:

$$X^i \to \frac{\tilde{x}^i}{1 - \tilde{x}^{d+1}} . \tag{14.8}$$

The utility of these coordinates becomes obvious when one considers the action of the ambient Lorentz transformations $x^\mu \to \Lambda^\mu_\nu x^\nu$ on the new coordinates. Indeed, near the conformal boundary, it can be shown that $x^- \to x^-$ and that the X^i transform under the full conformal group.

This last property allows us to define a scale-invariant theory holographically using the action principle for massive particles following bulk geodesics. To see how this can be done, consider the action for a single particle of mass m following a geodesic in dS

$$S(X^i_{in}, X^i_{out}) = \lim_{t_0 \to \infty} \int_{-t_0}^{t_0} dt \left[m\sqrt{-\eta_{\mu\nu}\dot{x}^\mu \dot{x}^\nu} + \lambda \left(\eta_{\mu\nu} x^\mu x^\nu - \ell^2 \right) \right] , \tag{14.9}$$

where X^i_{in} and X^i_{out} are the asymptotic values of the coordinates X^i on the past and future conformal boundary. The Lagrange multiplier λ enforces the constraint keeping the particle on the dS hyperboloid. If we evaluate this along the classical solution while carefully taking the limit, S becomes of a function of the asymptotic values of X^i. Moreover, as was just indicated, it is also conformally invariant. This means that it can be interpreted as the Hamilton–Jacobi function of some holographically defined conformally invariant theory.

In [6], S is explicitly computed in this limit. The result is

$$S = \frac{m\ell}{2}\left[\ln\left(\frac{(X_{in} - X_{out})^2}{\epsilon^2} - 2\right) + \mathcal{O}(\epsilon^4)\right], \tag{14.10}$$

where $\epsilon = \ell/t \to 0$ as $t \to \infty$. This behaves exactly like the Hamilton–Jacobi functional of a reparametrization invariant theory with potential equal to $V = \frac{1}{X^2}$, which is well-known to be scale invariant. We see that a free massive particle in dS spacetime can be equivalently described by a scale-invariant particle in a reparametrization invariant theory. Furthermore, the bulk dS isometries map explicitly to conformal transformations in the dual theory, as advertised.

References

1. H. Gomes, S. Gryb, T. Koslowski, Einstein gravity as a 3D conformally invariant theory. Class. Quant. Gravity **28**, 045005 (2011), arxiv:1010.2481 [gr-qc]
2. H. Weyl, Reine infinitesimalgeometrie. Math. Z. **2** (1918)
3. G.t. Hooft, The conformal constraint in canonical quantum gravity. arxiv:1011.0061 [gr-qc]
4. P.D. Mannheim, Making the case for conformal gravity. Found. Phys. **42** 388–420 (2012). arxiv:1101.2186 [hep-th]
5. R. Percacci, Asymptotic safety. arxiv:0709.3851 [hep-th]
6. S. Gryb, Observing shape in spacetime. (In preperation) (2014)
7. D.K. Wise, Holographic special relativity. arxiv:1305.3258 [hep-th]
8. R.W. Sharpe, *Differential Geometry: Cartan's Generalization of Klein's Erlangen Program* (Springer, New York, 1997)
9. S. Gryb, F. Mercati, 2+1 gravity on the conformal sphere. Phys. Rev. D **87**, 064006 (2013). arxiv:1209.4858
10. D. Arnold, J. Rogness, Mobius transformations revealed. http://www.ima.umn.edu/~arnold// moebius/

Chapter 15
Without Cause

Mark Feeley

> *Some mon just deal wit' information. An' some mon, 'im deal wit' de concept of truth. An' den some mon deal wit' magic.*
> —Nernenny, Rastafarian "Bush Doctor" (Nernenny, quoted in [1], p. 1)

Abstract Physicists increasingly accept that information is more fundamental than material things, but if material things are not fundamental, then neither are material causes: we will live in a world without cause. We thus examine the steps and missteps by which information came to be seen as more fundamental, examine the flaws and risks of a purely informational view, and consider a possible approach to restoring a belief in material things and material causes.

Introduction

It will come as a surprise to most of the general public, and even to most beginning students of physics, that a great many theoretical physicists believe in magic and not physical law. Guided by the dogma of quantum theory, many (and perhaps most) physicists accept that in the so-called quantum world, events can happen with no natural cause at all: a particle decays into other particles, particles are detected here versus there, or a spin is resolved as up or down. In the orthodox quantum view, the outcomes in these examples are said to be defined at the instant of measurement as the result of some sort of stochastic process. Unfortunately, though the term "stochastic process" has a pleasantly scientific tone, if there is no natural cause for such events, then we can safely replace this term with "supernatural cause" or "magic" without any change in meaning. Even if most physicists do not admit to a general belief in magic, they must admit that there is a general loss of faith in natural causes which

M. Feeley (✉)
26 Marchbrook Circle, Kanata, ON K2W 1A1, Canada
e-mail: mfeeley01@gmail.com

A. Aguirre et al. (eds.), *It From Bit or Bit From It?*,
The Frontiers Collection, DOI 10.1007/978-3-319-12946-4_15

has caused the search for physical law to be almost abandoned. No one now seeks to understand *why* a particle decays at a given time or a spin resolves as up or down: very few even believe there is a why. Expressed in John Wheeler's terms, physicists no longer believe in "It".

The modern idea is that we live in an informational world, not a physical one, and that the fundamental laws that we should seek are the laws of information, not physical laws. This is misguided and it is dangerous. Information theory necessarily augments physical theory when knowledge is limited, but cannot replace it. The choice facing physics is not one of information theory versus physical theory, it is information theory plus physical theory versus information theory plus magic. That physicists do not believe in a *physis*, a physical world governed by physical law, can only be seen as a crisis for physics. However, with some intellectual discipline and some retraining, we might escape this crisis.

Into the Crisis: Steps and Missteps

To understand the way out of the crisis, we must first understand the way in. Ernst Mach laid the groundwork for the rise of the informational view by arguing that physical science could not aspire to be a true description of reality, but should instead be the best summary of the available facts about reality:

> The goal which it (physical science) sets for itself is the simplest and most economical abstract expression of facts.[1]

By introducing the critical distinction between "facts about reality" and "reality itself", Mach allows the facts and the reality to diverge, and thus admits into physical theory two elements that are now essential: that the facts may be observer dependent and that the facts may be limited. Of course Einstein used the observer dependence of facts with stunning success in his developments of both relativity theories. The second element, the limitation of information, generally necessitates the use of probabilistic or statistical treatments, and is a key feature of both statistical mechanics and quantum mechanics. Claude Shannon created modern information theory in 1948 as a development of standard probability theory, and a few years later, Edwin Jaynes pioneered the application of information theory in physics. Jaynes redeveloped much of statistical mechanics in terms of Shannon's information theory, and gave us a new understanding of entropy as an informational or epistemic concept rather than a thermodynamic one. However, the uses of information theory in quantum and statistical mechanics can be sharply contrasted. Statistical mechanics acknowledges the limitation of knowledge without denying the existence of an underlying reality; the physical view of reality is still unquestionably more fundamental than the informational view. Furthermore, the underlying model is deterministic—no intrinsic randomness is assumed—and the physical model actually informs the statistical

[1] E. Mach, "The Economical Nature of Physical Inquiry", excerpted in [2].

model. On the other hand, depending on the interpretational flavour, quantum theory is either indecisive about the nature and existence of reality or denies reality outright. It is only with quantum theory that an informational view begins to be considered as somehow more fundamental than a physical view. However, the confusion and doubt about reality in quantum theory, and thus the support for the primacy of an informational view, stems from a misunderstanding: the creators of quantum theory simply did not understand or apply probability theory correctly.

The Way Out: Relearning Probability and Quantum Theory

Those of us who struggled with discipline in our early school years may recall "writing lines" as punishment for our misdemeanours: 500 lines of "*I will not chew gum in class*..." and so on. This type of remediation would be of great benefit to physicists today. Understanding of physics would be vastly improved if in the first class of every course in quantum theory—beginning and advanced, undergraduate and graduate—students were handed a stack of foolscap paper and assigned to repeatedly write a version of the extraordinarily lucid and pointed line given to us by de Finetti[2]:

Probability is not real.
Probability is not real.
Probability is not real...

In the second class, students could discuss at length: if probability is not real, then what is it exactly? The familiar game of heads and tails, coin tossing, tells us all that we will need. Students can be asked to consider a coin tossing experiment in which they are given a coin tossing machine of such precision that a coin initially placed in the machine with a given face up will land upon a table with the same face up with certainty. The table is glass so that the coin may be read by an observer above or below the table. Before any toss, information will be given, stipulating whether the coin is placed in the machine with heads up/down/unknown and whether the observer determines the outcome by reading the coin from above/below/unknown. They will be asked to understand the meaning of the term "the probability of heads or tails given this information", denoted $P(H|I)$ and $P(T|I)$.

First, they must decide what heads and tails actually are. Of course, coins have symmetries and physical features, namely two faces embossed with pictures, and we can arbitrarily label each face as heads or tails. However, by "the probability of heads" we mean something like "the probability that the face arbitrarily labelled heads is visible to an observer at the end of the experiment". In this context, heads is the name for a *state*, not a face. But a state of what? The first guess may be that heads and tails are final states of the coin, but brighter students will object that they cannot simply be states of the coin, since the reading also depends on the position of the

[2] Original quote: "Probability does not exist", de Finetti [3].

observer. Furthermore, heads and tails have no meaning at all until the experiment is complete, so coins cannot be said to "have" heads or tails states. They will conclude, hopefully, that heads and tails are *outcomes of the experiment*. They are not properties or states of the coin; they are states of the outcome. The difference is profound. As we allow no other outcomes, they will agree that the set {*heads, tails*} is the entire outcome space of the experiment.

The students will then be asked to explain how probability is determined and what variables affect it. They will consider the probabilities that would seem reasonable to them depending on whether they are told the coin is initially oriented with heads up/down/unknown, supposing in all three cases that they are told the position of the observer (above). Knowing the precision of the machine, they will decide: if the coin is known to be initially oriented with heads up then we should reasonably assign a probability of to the heads outcome, $P(H|I) = 1$, if known to be initially oriented with heads down then we would reasonably assign $P(H|I) = 0$, and if the initial orientation is unknown then we would assign $P(H|I) = 0.5$. From this they will easily conclude the key features of probabilities: probabilities are assigned *by us*, and probabilities are related to our information about the conditions of the tossing experiment and not directly to the coin, the tossing machine, or any other physical thing.

So, we can explain, to quantify the state of our knowledge using the methods of probability theory, we first assume an outcome space, and then we assign a probability to each outcome in that outcome space. In probability theory, this probability distribution over the outcome space entirely quantifies our knowledge. Probability is then used to make predictions and we can describe the methods of probability theory, statistics, or information theory as various types of epistemological calculus.

At this point, students should understand "Probability is not real". Probability is not itself physical and thus does not exist in space or time. Probability is not a property of, or directly associated with, either physical things or physical systems, or even states of physical things or systems. There is no such thing as a "physical probability". Probability is an *epistemological* measure assigned *by us* to *outcomes* of experiments, and is used to quantify our knowledge in an epistemological calculus.

The third class could entail some more advanced lines:

Whenever I see probability in an expression, I will interpret the expression as epistemological, not ontological.
Whenever I see probability in an expression, I will interpret the expression as epistemological, not ontological...

In plain English: "If a theory or an expression has a probability in it, then it isn't about something physical, it's about outcomes we think might occur".

After this rote training and discussion, it might then be safe to introduce quantum theory. We will attempt to present the theory without the history, the mystery, the philosophies, or the interpretations. We will hope that students have not been tainted by too much prior exposure to the theory—in popular science books, FQXi contests, and so on. We will give them only equations and some guidance regarding the notation, and ask them to deduce what they can of the meaning of the equations and the theory. We need only tell them that the theory features expectation values, and even the least astute will recognize that they are dealing with a probabilistic theory: an epistemological theory, not an ontological one. They will expect a theory of experiments and outcomes.

Since they expect experiments, we first develop a classical theory of experiments, once again referring to the coin tossing experiment C. As heads and tails are outcomes rather than mathematical entities, we must first choose a mathematical representation for these states. We will choose to make use of vector algebra, and so will represent outcomes as "directions" in the outcome space. Thus, we define $|H\rangle$ as a unit vector in the "heads direction" ($\widehat{H}$ or $\vec{H}$ would be more obvious vector notations, but the bra-ket notation will be useful later). We then want to assign two different quantities to an outcome state: a value of some kind (usually numeric) and a probability. The values we assign for head and tails are denoted c_H and c_T. The choice is entirely arbitrary, but $c_H = 1$ and $c_T = -1$ would be a typical choice.

To represent this value assignment mathematically, we define an operator $\tilde{C}$ satisfying

$$\tilde{C}|H\rangle = c_H|H\rangle, \quad \tilde{C}|T\rangle = c_T|T\rangle.$$

Thus, $\tilde{C}$ simply represents the process of assigning a numeric value c_H to the abstract outcome state $|H\rangle$. The key benefit of this operator representation is that it clearly distinguishes between the value we have assigned to an outcome state and the outcome state itself. We call $|H\rangle$ and $|T\rangle$ the eigenvectors of $\tilde{C}$ and call $c_H$and c_T the eigenvalues of $\tilde{C}$. We require that our operator $\tilde{C}$ generates values for all possible outcomes of the experiment, and formally this requires that the eigenvectors of $\tilde{C}$ span the outcome space. We will further specify that the outcomes are defined by orthonormal vectors, so that:

$$\langle H|H\rangle = \langle T|T\rangle = 1, \quad \langle H|T\rangle = \langle T|H\rangle = 0.$$

If $|H\rangle$ and $|T\rangle$ are defined as orthonormal vectors spanning the outcome space, then any vector $|\psi\rangle$ in the outcome space of the experiment can be written

$$|\psi\rangle = h|H + t|T\rangle.$$

We next assign probabilities $P(H|I)$ and $P(T|I)$ to the outcome states, and according to probability theory, the expectation value for a coin toss $\langle C\rangle$ is

$$\langle C\rangle = c_H P(H|I) + c_T P(T|I).$$

This is all we need for many purposes, but to capture the full representation of the coin toss experiment, and recognizing that the process of assigning values was arbitrary, we may wish to express $\langle C\rangle$ in terms of an outcome vector $|\psi\rangle$ and the operator $\tilde{C}$ explicitly. To do this, we define quantities ϕ_H and ϕ_T, called *probability amplitudes*, which are complex roots of the probabilities, satisfying $\phi_H{}^* \phi_H =$

$P(H|I)$ and $\phi_T{}^* \phi_T = P(T|I)$. We will ignore the detail of why we choose complex versus real roots in this essay, and indeed the phases will rarely matter. We then further define our outcome vector to be

$$|\psi\rangle = \phi_H|H\rangle + \phi_T|T\rangle,$$

so that

$$\tilde{C}|\psi\rangle = c_H\phi_H|H\rangle + c_T\phi_T|T\rangle.$$

Standard vector algebra defines the scalar product of vectors (with complex components) as

$$\langle\psi_2|\psi_1\rangle = h_2{}^* h_1 + t_2{}^* t_1,$$

and we can use this scalar product to link the operator and outcome vector representations to probability theory and give us $\langle C\rangle$ in terms of $\tilde{C}$ and ψ,

$$\begin{aligned}\langle C\rangle = \langle\psi|\tilde{C}|\psi\rangle = \langle\psi|\tilde{C}\psi\rangle &= \phi_H{}^* (c_H\phi_H) + \phi_T{}^* (c_T\phi_T) \\ &= c_H P(H|I) + c_T P(T|I).\end{aligned}$$

We now have a complete representation of our coin toss experiment, which allows us to capture outcome states ($|H\rangle$), the probabilities ($P(H|I)$) and probability amplitudes (ϕ_H) assigned to those states, the values (c_H) assigned to those states, and the procedure ($\tilde{C}$) which assigns those values. This theory of experiments is simply standard probability theory combined with an operator representation of experiments and some vector algebra. We can easily generalize to more outcomes or to continuous outcomes. Although it should go without saying, coin tossing is perfectly classical.

Now, finally, we can begin quantum theory. Students can be grandly told their first "fundamental postulate of quantum mechanics":

> With any observable A, we associate an operator $\tilde{A}$ it which acts on ψ, and the only results of a measurement of A will be one of the eigenvalues a_i of $\tilde{A}$, satisfying $\tilde{A}\psi_i = a_i\psi_i$.

Having just seen this in the context of coin tossing, this will seem blasé. They will not view this as having the exalted status of a postulate or even in any way quantum, just the standard structure of a theory of experiments—classical, quantum, or otherwise. They will instantly recognize that an observable is a name for a type of *experiment* (coin tossing), not a property of some physical thing, that a measurement is an instance of that experiment producing a single outcome (a toss), that the eigenvectors ψ_i are vector representations of the outcomes (*heads* or *tails*), and that the set of eigenvectors $\{\psi_i\}$ is the representation of the full outcome space of the experiment ($\{$*heads, tails*$\}$), and the eigenvalues a_i are values $(1, -1)$ which we have chosen to assign to outcomes.

Given the "fundamental postulate" for expectation values

$$\langle A\rangle = \langle\psi|\tilde{A}|\psi\rangle = \Sigma\langle\phi_i\psi_i|a_i|\phi_i\psi_i\rangle = \Sigma a_i\phi_i{}^*\phi_i = \Sigma a_i P(i),$$

they will recognize a_i as the value assigned to the ith outcome of experiment A, and $P(i)$ as the probability of the ith outcome. They will see a straightforward probability-based theory of experiments.

The meaning of the wavefunction ψ would be quite mundane and just as it was for coin tossing: it is a vector in the outcome space of an experiment. The wavefunction captures both the outcome states (defined as part of the definition of the experiment) and the probabilities which *we have assigned* to those states. Not wishing to write more lines, they will not accept that the wavefunction is in any way physical, or even directly associated with anything physical. Probability is not real.

If you were to tell them, with fanfare befitting such a great mystery, that the function ψ "collapses" upon measurement, instantaneously and everywhere, they would be astonished only at your theatrics, as it is quite obvious that the probability of an outcome becomes 1 when that outcome is known.

We can then ask them to attempt to determine, from equations alone, what sort of physical thing the theory might describe. Of course, they will accept that there is some physical thing, as they have no particular reason to suspect otherwise. We can tell them that we seem to get the best predictions in many experiments if we assume that the function ψ evolves according to wave-like equations such as

$$i\partial_t\psi = (-\frac{1}{2m}\nabla^2 + V)\psi \quad or \quad (\partial_t^2 - \nabla^2)\psi = -m^2\psi$$

But why should *we choose to assign* a spatially and temporally varying probability to an outcome? Since any information is given at the start of the experiment and does not change, *we must have reason to believe* that some feature of the physical thing in our experiment evolves in a wave-like fashion. For example, suppose that our experiment (our observable) is named "water heights on vertical sticks in the Bay of Fundy", our outcomes are heights h, our outcome space is the continuous domain $[h_{lo}, h_{hi}]$, and our information I is that the Bay of Fundy is famous for its tides, that tides have exhibited periodic behaviour with period ≈12–24 h in all previously known cases, and we are given $h(t_o)$. Given I, we have sound reasons to believe that we should assign a time-varying probability to any outcome . If $h \approx h_{hi}$ now, we should assign a high probability that $h < h_{hi}$ in about 6 h. With more information, such as a physical theory of tides and positions of the sun and moon, we could refine our probability model further and would presumably be led to assign probabilities which vary in a wave-like fashion in time. The equations of quantum theory do not cast doubt on the existence of an underlying physical world. On the contrary, the evidence offered by the equations positively suggests a physical world with wave-like features. Students would be keen to understand this physical world.

Unfortunately, students are not taught quantum theory this way. Instruction often begins with tales of the supposed failings of "classical physics", and proceeds to such incomprehensible pronouncements as "we may associate a wave function with every particle, and the wave function is a complex probability amplitude whose squared modulus represents the probability of finding the particle at a point in space-time", "the wavefunction captures all that we may know about *the system*", or "the wavefunction is a probability distribution over the state space of *the system*". These ideas are nonsense, as they all clearly associate the probability with the thing—the particle or the system—not the outcome of an experiment given some information. The faulty notion of physical probabilities is intrinsically assumed. The idea of physical probabilities pervades quantum theory, and we might hear "the electron has a probability of 1/2 of being found in a state of spin up or down in any direction". An electron need not "have" a spin state at all: a coin toss experiment has a heads state, but a coin does not, and a spin experiment has a $\uparrow$ state, but an electron may not. An electron has features which can be manipulated in experiments with outcome states labelled $\uparrow$ and $\downarrow$, we can say no more. We are told the Uncertainty Principle relates two properties of a particle, but it cannot: it relates only the outcomes of two correlated experiments.

Students taught "our way", with a clear idea of what a probability is, would reject all of these assertions out of hand. They would know that probabilities are not real. Our students would demand to know the information which was given in order to assign a probability to an outcome of a given experiment. They would not doubt the existence of a physical thing upon which they experiment, and they would demand to know of any physical models which they might use to better estimate probabilities. They would recognize that they have limited information, but would not doubt the existence of causes. In short, they would not be deceived.

"It from Bit": Here Be Dragons

Unfortunately, most people have been deceived. The flawed concept of physical probabilities is almost inextricably tied into the foundations of quantum theory and leads directly to most of the confusion in physics. It leads to all of the confusion about epistemology and ontology, about information and reality, and it leads, inevitably, to "It from Bit". Now, John Wheeler has made very many important contributions to physics, but "It from Bit" is simply not his finest hour. In fairness, the essential idea behind "It from Bit" is not even his, as in 480 BC the Milesian Greek philosopher Anaxagoras taught that all things are created by the mind. Wheeler's principal innovation over Anaxagoras was to assert that the information received by the mind is digital, which indeed may not be trivial as a form of limitation of information, but the central idea remains the same. However, "It from Bit" effectively captures the zeitgeist of post-quantum physics, and as a quote, has a mystical, Zen-like quality which gives it great power. Unfortunately, it is both wrong and unhelpful.

To see why Wheeler's "It from Bit" is wrong, we can examine his own explanation:

> Otherwise put, every "it"—every particle, every field of force, even the space-time continuum itself—derives its function, its meaning, its very existence entirely—even if in some contexts indirectly—from the apparatus-elicited answers to yes-or-no questions, binary choices, bits. "It from Bit" symbolizes the idea that every item of the physical world has at bottom—a very deep bottom, in most instances—an immaterial source and explanation; that which we

> call reality arises in the last analysis from the posing of yes-no questions and the registering of equipment-evoked responses; in short, that all things physical are information-theoretic in origin and that this is a participatory universe.[3]

This passage is not terribly lucid, but we can attempt to parse it. Wheeler defines "Bit" as a set of apparatus-elicited answers to yes or no questions. To make any logical sense of the explanation, we must ask two questions: "Is the apparatus It or Bit?" and "Questions about what?" The first question actually has no satisfactory answer. If the apparatus is "Bit" then we unfortunately are not left with any "It", all is "Bit". Such a position may well be valid, but anyone seriously holding this opinion is best advised to abandon physics and pursue philosophy or psychology. On the other hand, if the apparatus is "It", then following Wheeler, this apparatus owes its existence to yes or no questions asked by another apparatus, and that one by yet another, in infinite regress. This is nothing more than a fancy proposal for implementing the famous "turtles all the way down", and is really not very useful. Sadly, turtles that ask yes or no questions are no more believable as a basis for reality than regular turtles.

As to the second question, unless the apparatus concocts answers of its own volition, then we must presume that the apparatus gives its yes or no answers based on what it can determine of some domain external to itself. Even from Wheeler's own definition of "Bit", it is thus apparent that information is *information about something*, not information about information or information in the abstract. "Bit" is about "It". As should have been quite obvious at the outset (except apparently, to theoretical physicists), the information is derived from the something, not the something from the information. Despite the imaginative sophistry, "It from Bit" lacks any logical consistency and really does not pass muster.

Now, some ideas turn out to be illogical or wrong but nevertheless useful, and Newton's theory of gravity as instantaneous force-at-a-distance is one such example. Instantaneous force-at-a-distance is now thought to be wrong, but the idea remains computationally useful in many domains and was also very useful as a stepping stone in the search for better physical law. Wheeler's idea has no such merits.

"It from Bit" suggests that we should consider information theory, not physical theory, as fundamental. Whenever we have limited information, some form of probabilistic or informational theory actually *must* be used. However, these methods must only be used to augment a physical theory. Whatever the many uses and merits of information theory, it is fundamentally empiricist, and does not require or seek causes, mechanisms, explanations, or physical laws. Thus, the view that physics is information theory implicitly suggests that we can abandon the necessary search for those mechanisms, causes and laws.

Jaynes, on the other hand, was always particularly careful to distinguish between reality and what we know about reality. He even gives a name to the mistaken assumption that what we know about reality *is* reality: he calls this the Mind Projection Fallacy. The principal danger of the Mind Projection Fallacy is the denial of causes: "I do not know the cause" therefore "there is no cause". Unfortunately,

[3] J. Wheeler, in [4].

by claiming that reality "owes its very existence to" what we know about reality, Wheeler became the poster boy for the Mind Projection Fallacy. Jaynes describes the danger of a loss of a faith in physical causes very well:

> In current quantum theory, probabilities express our own ignorance due to our failure to search for the real causes of physical phenomena—and worse, our failure even to think seriously about the problem. This ignorance may be unavoidable in practice, but in our present state of knowledge we do not know whether it is unavoidable in principle, the "central dogma" simply asserts this, and draws the conclusion that belief in causes, and searching for them, is philosophically naïve. If everybody accepted this and abided by it, no further advance in understanding of physical law would ever be made; indeed, no such advance has been made since the 1927 Solvay congress in which this mentality became codified into physics. But it seems to us that this attitude places a premium on stupidity; to lack the ingenuity to think of a rational physical explanation is to support the supernatural view.[4]

Of course Jaynes, of all people, is not arguing that we should not use statistical methods, only that statistical methods should augment but never replace the search for physical theory. Recent work by a number of physicists on reconstruction of quantum theory with a new a set of informational axioms is valuable and will help provide clarity on the nature of quantum theory to be sure, but will not and cannot produce a physical theory. Indeed, informed as it is by the "It from Bit" philosophy, such work does not even strive to do so. A physical theory underlying quantum theory is also needed, and it is most definitely not naïve to pursue it.

Conclusion

"What is the relationship between epistemology and ontology, mind and matter, information and reality, or Bit and It?" However we phrase it, the question is a very old one—older than Wheeler and Jaynes, older than Mach, older than Anaxagoras, and possibly older even than the cave painters of Lascaux. Just as the ancient painters used pigments to create representations of the reality they saw, we use mathematics to create representations of the reality we see. To be sure though, there is a reality, "It", and the information creates a representation of that reality. Our information is very likely constrained to be digital as Wheeler suggests, thus "Bit", and may be limited in other ways, but "It" does not derive from "Bit". "Bit" manifestly derives from "It".

Since information about reality is necessarily limited, physicists can and must make use of information theory to understand physics, but it certainly does not follow that information must be seen as fundamental. Information theory is a tool which allows us to quantify and best use our knowledge about the physical world in a concise mathematical form. Physicists should use this tool, but the task of physicists is nothing other than to discover physical theory, physical law and physical causes. The task cannot be avoided or shirked from, and cannot be wished away with information

[4] E. Jaynes, in [5], p.1013.

theory, stochastic processes, or mystical incantations. Quantum theory, the source of the problem, will have to be rethought and relearned. Since information about reality and reality itself are different things, they must be differentiated in any theory. Quantum theory provides no such distinction, and thus, despite its predictive value, must be wrong. Quantum theory must be reworked or replaced with a theory which provides the same results, but which offers a clear epistemological/ontological (or Bit/It) boundary, or it is unlikely that progress can be made.

Probability is not real, but causes are real. We must not believe in magic. We can be optimistic that a physical theory underlying quantum theory can be found—that "It" can be restored to primacy. Indeed, it is Wheeler himself who best inspires us to continue the search:

> Behind it all is surely an idea so simple, so beautiful, that when we grasp it—in a decade, a century, or a millennium—we will all say to each other, how could it have been otherwise?
>
> – John Archibald Wheeler[5]

References

1. T. White, *Catch a Fire—The Life of Bob Marley* (St. Martin's Griffin, New York, 2006)
2. J. J. Kockelmans *Philosophy of Science: The Historical Background* (The Free Press, New York, 1968)
3. B. de Finetti, in *The Theory of Probability: A Critical Introductory Treatment*, vol. 2, ed. by A. Lachi, A. Smith (Wiley, London, 1990)
4. J.A. Wheeler, Complexity, entropy, and the physics of information, in *Information, Physics, Quantum: The Search for Links*, ed. by H. Wojcicch, H. Zurek (Westview Press, 1990)
5. E.T. Jaynes, *Probability Theory: The Logic of Science*, ed. by L. Bretthorst (Cambridge University Press, 2003)
6. J.A. Wheeler, How come the quantum? Ann. New York Acad. Sci. **480** (1986)

[5] J. Wheeler, in [6], p. 304.

Chapter 16
Reality, No Matter How You Slice It

Ken Wharton

Abstract In order to reject the notion that information is always *about something*, the "It from Bit" idea relies on the nonexistence of a realistic framework that might underly quantum theory. This essay develops the case that there *is* a plausible underlying reality: one actual spacetime-based history, although with behavior that appears strange when analyzed dynamically (one time-slice at a time). By using a simple model with *no* dynamical laws, it becomes evident that this behavior is actually quite natural when analyzed "all-at-once" (as in classical statistical mechanics). The "It from Bit" argument against a spacetime-based reality must then somehow defend the importance of dynamical laws, even as it denies a reality on which such fundamental laws could operate.

Introduction

Information, not so long ago, used to always mean knowledge *about something*. Even today, under layers of abstraction, that's still the usual meaning.[1] Sure, an agent can be informed of a string of bits (via some signal) without knowing what the bits refer to, but at minimum the agent has been informed about the physical signal itself.

Quantum theory, however, has led many to question this once-obvious connection between knowledge/information and an underlying reality. Not only is our information about a quantum system indistinguishable from our best physical description, but we have failed to come up with a realistic account of what might be going on

[1] The technical concept of Shannon Information is distinct from this everyday meaning, although they are often erroneously conflated. Shannon Information is perhaps better termed "source compressibility" or "channel capacity" (in different contexts), and is a property of (real) sources or channels. [1] This essay utilizes the everyday meaning of "information": an agent's knowledge.

K. Wharton (✉)
Department of Physics and Astronomy, San José State University,
San José, CA 95192-0106, USA
e-mail: kenneth.wharton@sjsu.edu

A. Aguirre et al. (eds.), *It From Bit or Bit From It?*,
The Frontiers Collection, DOI 10.1007/978-3-319-12946-4_16

independent of our knowledge. This blurring between information and reality has led to a confusion as to which is more fundamental.

The remarkable "It from Bit" idea [2] that *information* is more fundamental than reality is motivated by standard quantum theory, but this is a bit suspicious. After all, there's a long "instrumentalist" tradition of only using what we can measure to describe quantum entities, rejecting outright any story of what might be happening when we're not looking. Using a theory that only comprises our knowledge of measurement outcomes to justify knowledge as fundamental is almost like wearing rose-tinted glasses to justify that the world is tinted red.

But any such argument quickly runs into the counterargument: "Then answer the question: What *is* the (objective) reality that our information of quantum systems is actually *about*?" Without an answer to this question (that differs from our original information), "It from Bit" proponents can perhaps claim to win the argument by default. The only proper rebuttal is to demonstrate that there is some plausible underlying reality, after all.

This is generally thought to be an impossible task, having been ruled out by various "no-go" theorems [3–5]. But such theorems are only as solid as their premises, and they all presume a particular sort of independence between the past and the future. This presumption may be valid in a universe that uses dynamical laws to evolve some initial state into future states, but there is a natural alternative to this dynamic viewpoint. As previously argued in [6] (and summarized in Appendix I), instead of the universe solving itself one time-slice at a time, it's possible that it only looks coherent when solved "all-at-once".

This essay aims to demonstrate how this all-at-once perspective naturally recasts our supposedly-complete information about quantum systems into *incomplete* information about an underlying, spacetime-based reality. After some motivation in the next section, a simple model will demonstrate how the all-at-once perspective works for purely spatial systems (without time). Then, applying the same perspective to spacetime systems will reveal a framework that can plausibly serve as a realistic explanation for quantum phenomena.

The result of this analysis will be to dramatically weaken the "It from Bit" idea, showing that it's possible to have an underlying reality, even in the case of quantum theory. We may still choose to reject this option, but the mere fact that it is on the table might encourage us *not* to redefine information as fundamental—especially as it becomes clear just how poorly-informed we actually are.

Instants Versus Spacetime

The case for discarding dynamics in favor of an all-at-once analysis is best made by analyzing quantum theory [6], but it's also possible to frame this argument using the *other* pillar of modern physics: Einstein's theory of relativity. The relevant insight is that there is no objective way to slice up spacetime into instants, so we must not assign fundamental significance to any particular slice.

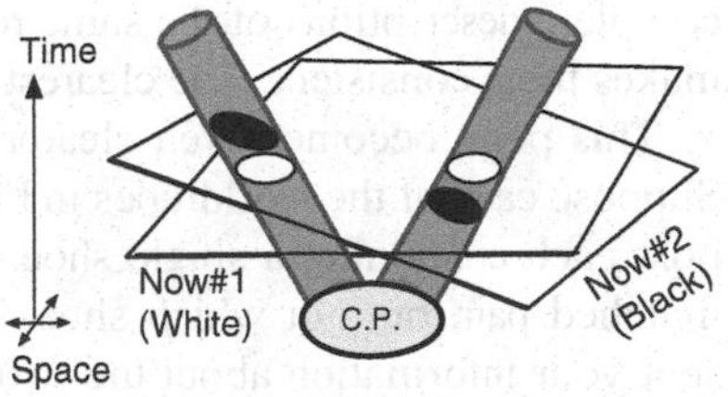

Fig. 16.1 A spacetime diagram, demonstrating the unreality of "now". (See text.)

Figure 16.1 is a standard spacetime diagram (with one dimension of space suppressed). If run forward in time like a movie, this diagram represents two spatial objects that begin at a common past (C.P.) and then move apart. But if viewed all-at-once, the figure instead shows two grey "worldtubes" that intersect in the past. In relativity, as we are about to see, it is best to analyze this picture all-at-once.

The most counter-intuitive feature of special relativity is that there is no objective "now". Simultaneous events for one observer are not simultaneous for another. No observer is right or wrong; "now" is merely subjective, not an element of reality. An illustration of this can be seen in Fig. 16.1. Observer #1 has a "now" that slices the worldtubes into two white ovals, while Observer #2 has a "now" that slices the worldtubes into two black ovals. Clearly, they disagree.

This fact implies that any dynamical movie made from a spacetime diagram will incorporate a subjective choice of how to slice it up. One way to purge this subjectivity is to simply view a spacetime diagram as a single 4D block. After all, with no objective "now", there is no objective line between the past and the future, meaning there can be no objective difference between them.

Such a claim is counter-intuitive, *but this is a central lesson of relativity.* The only difference between the future and the past, in this view, is subjective: we don't (yet) know any of our future. Arguments such as "But the future isn't real *now*" are no more meaningful than arguing "Over there isn't real right here".

A more reasonable fallback for the dynamicist is not to deny that spacetime *can* be viewed as a single 4D block, but rather to note that if dynamical equations govern the universe[2] then *any* complete spacelike slice suffices to generate the rest of the block (via dynamical equations). So while no one slice is special, they're all equally valid inputs from which the full universe can be recovered. Taken to an extreme, this viewpoint leads to the notion that the 4D block is filled with redundant permuted copies of the same 3D slice. It also forbids a number of solutions allowed by general relativity, spacetime geometries warped to such an extent that they only make sense all-at-once.

The other problem with this sliced perspective is that it all but gives up on objectivity. Even if it's *possible* to generate the block from a single slice (a point I'll dispute later on), how can one 3D slice truly generate the others if it is a subjective choice? In Fig. 16.1, if *both* the white ovals and the black ovals are different

[2] Along with other subtleties, such as the existence of Cauchy data.

complete descriptions of the same reality, it's the 4D worldtubes they generate that makes them consistent. The clearest objective reality requires a bigger picture.

This point becomes even clearer when one introduces (subjective) uncertainty. Suppose each of the worldtubes in Fig. 16.1 represent a (temporally extended) shoebox, each containing a single shoe. Also suppose that you knew the shoes were a matched pair, but not which shoe (R or L) was in which box (1 or 2). To represent your information about the two boxes after they had separated (say, the white ovals in Fig. 16.1), you might use an equal-probability mix of both possibilities: $S_{mix} = [50\,\%(L_1R_2), 50\,\%(L_2R_1)]$. This is not a potential state of reality, but a state in a larger "configuration space" that weights possibilities that *do* fit in spacetime. Note that having less knowledge forces a more complicated description, even if the underlying reality is assuredly either L_1R_2 or L_2R_1.

For these restricted-knowledge situations, the all-at-once viewpoint is invaluable if we are to make sense of what is going on when we open a shoebox and learn which shoe is inside. If we take the dynamic view that we need only keep track of the 3D white ovals in Fig. 16.1 to describe the entire 4D system, then S_{mix} might seem to give us everything we need; from it we can compute outcome probabilities and the correlations between the two boxes. Upon learning that (say) the left shoe is in box 1, we can even update our knowledge of the 3D state S to the appropriate $[100\,\%(L_1R_2)]$. But what is lost in this viewpoint is the mechanism for the updating; if our entire description is that of the 3D white ovals, this updating process might appear nonlocal, as if some spooky influence at box 1 is influencing the reality over at box 2.

Sure, we know that nothing spooky is going on in the case of shoes, but that's only because we already know there's an underlying reality of which S_{mix} represents (subjective) *information*. If the existence of an underlying reality is doubt (as in quantum theory), then analysis of the 3D state S_{mix} cannot address whether anything spooky is happening. To resolve that question, one has to look at the entire 4D structure. All at once.

In the all-at-once viewpoint, after finding the left shoe in box 1 we update our local knowledge to L_1 (updating occurs when we learn new information). But thinking in 4D, we also update our knowledge of the *past*; we now know that that L_1 back in the C.P. This in turn implies R_2 back in the C.P, and this allows us to update our knowledge of R_2 in the present. It's the continuous link, via the past, that proves that we did not change the contents of box 2; it contained the right shoe all along. Throw away the analysis of the 4D link, and there's no way to be sure.

Before moving on, it's worth noting that this classical story cannot explain all quantum correlations; in fact, it's exactly the story ruled out by a no-go theorem [3]. Such theorems generally start from the classical premise that we can assign subjective probabilities p_i to possible 3D realities, W_i. States of classical information then naturally take the form $S = [p_1(W_1), p_2(W_2), \ldots, p_N(W_N)]$, a function on $3N$-dimensional configuration space. (Note the probabilities are all subjective; only one particular W is real; the rest are not.) The quantum no-go theorems have proven that such a state cannot explain quantum measurements without some classically-impossible feature, such as negative probabilities or faster-than-light signalling.

The standard thinking is that since any workable version of S cannot be classical information, it must be a new kind of reality in its own right. Effectively, the standard view[3] extends reality from spacetime to configuration space. But an alternative option, explored below, is that reality merely requires an extension from 3D to 4D, along with an all-at-once analysis. At this point it's probably not obvious how anything might change if the W's spanned 4D spacetime, but that's because the standard dynamical viewpoint makes any such extension trivial. (Thanks to dynamics, all the interesting information is always encoded in a 3D slice). Exploring this option therefore requires jettisoning dynamics.

Still, old habits die hard; it's difficult to think about time without also thinking in terms of dynamical equations, or the "Newtonian Schema" described in Appendix I. The 4D block is a good start, but it's time to demonstrate how it can be used to make physical predictions. Fortunately, it's a standard procedure, so straightforward that it's nearly trivial.

A Dynamics-Free Model

Physicists know how to do physics without dynamics, because we can analyze 3D systems for which there are no dynamics, by definition. A particularly useful approach is found in classical statistical mechanics, because in that case we never know the exact microscopic details, allowing us to deal with restricted knowledge situations.

The basic approach works like this. First, determine the possible underlying realities; call each one a "microstate" W_i. The key next step[4] is to assign each W_i an equal a priori probability, p_i. (Initially treat all possible states as equally likely.) If we learn new information—say, that W_9 is ruled out—we set $p_9 = 0$ and renormalize the remaining probabilities such that they sum to 1. Finally, we can determine the probability that the system has any particular feature by simply adding the probabilities of the microstates with that feature. One *could* introduce dynamics on top of this framework, but it's not a logical necessity.

For a simple example that will prove particularly relevant to quantum theory, consider Fig. 16.2. Each circle (perhaps a coin) can be in the state heads (H) or tails (T), and every line connects two circles. Each line has one of three internal colors; red, green or blue, but these colors are *unobservable* (they can sometimes be deduced, but not directly measured). The model's only "law" is that red lines must connect opposite-state circles ($H-T$ or $T-H$), while blue and green lines must connect similar-state circles ($H-H$ or $T-T$).[5]

Consider the following puzzle in the statistical mechanics framework:In Fig. 16.2a, if it is known that the bottom circle is H, what is the probability that the two circles in the dotted box are in the same state? It's easy enough to work out (see

[3] Including both deBroglie-Bohm [7] and Everettian [8] approaches.

[4] Sometimes known as the "fundamental postulate of statistical mechanics".

[5] This is effectively a much-simplified version of the Ising Model; see [9].

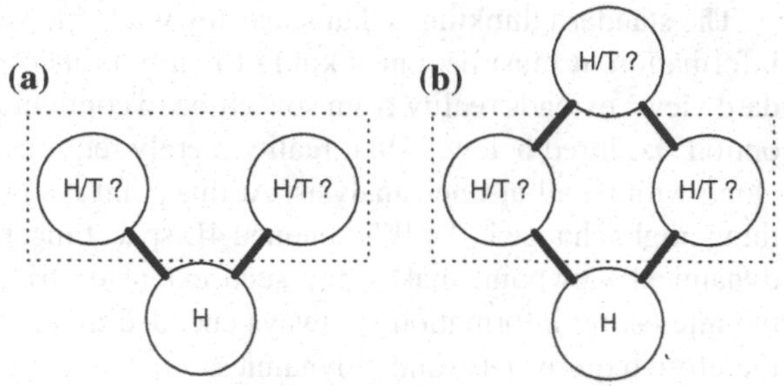

Fig. 16.2 Two geometries of a model, in which each circle can be Heads (H) or Tails (T). There are two line colors that connect matching circles (HH, TT) and a third line color to connect opposite circles (HT, TH). The interesting case is where one does not know whether the geometry is that of 2a or 2b

Appendix III for details) that there are four microstates where these two circles are HH, 2 microstates for HT, 2 for TH, and only 1 for TT. By assigning each of these nine states an equal probability, it should be evident that there is a 5/9 chance those two circles are the same, and a 4/9 chance that they're different.

For Fig. 16.2b, the same puzzle is trickier because now there's a fourth circle. In this case, the same style of analysis (also in Appendix III) reveals that the different geometry changes the probabilities. In place of a 5:4 probability ratio, here one finds a 25:16 ratio.

The most interesting example is a further restriction where one *does not know* whether the actual geometry is that of Fig. 16.2a or 16.2b. Specifically, one knows that the bottom circle is heads, and that the next two circles are connected, but not whether a fourth circle is connected (2b), or whether it is not (2a).

This is not to say there is no fact of the matter; there *is* some particular geometry—it's just unknown. This is not quite the same as the unknown circles or links (which also have some particular state), because this model provides no clues as to how to calculate the probability of a *geometry*. All allowable states may be equally likely, but that doesn't help us if we don't know which states are allowable in the first place. With this further-restricted knowledge, we would most naturally use an even higher-level configuration space to describe the probability of similar states in the dotted box, something like: $S_? =$ [If 2a then 5/9; If 2b then 25/(25 + 16)].

The next section will explore a crucial mistake that would lead one to conclude that *no underlying reality exists* for this statistical-mechanics-based model, despite the fact that an underlying reality does indeed exist (by construction). Then, by applying the above logic to a dynamics-free scenario in space *and* time, we'll see how we are making this same mistake in quantum theory.

Implications of the Model

The Independence Fallacy

Given the previous model, one might reasonably want to analyze a "slice" of the system (the dotted box) independent from the rest. But this can only be done by expanding the description of the box such that it includes *each* possible external geometry—effectively removing the "ifs" from $S_?$. Then, one can later use the actual geometry to extract the appropriate probability from the larger state space (either 5/9 or 25/41).

But this new perspective becomes quite mistaken if one further demands that the state of reality in the dotted box must *be* independent of the external geometry. It's obviously not true for this model, but given such an "Independence Fallacy" one would be led to some interesting conclusions. Namely, this all-possible-geometry state space would seem to be irreducible to a classical probability distribution over realistic microstates.

Given the Independence Fallacy, the argument would go like this: Geometry 2a implies a 5/9 probability of similar circles, while geometry 2b implies a 25/41 probability. But since the state must be independent of the geometry, the question "Are the two circles the same?" cannot be assigned a coherent probability. And if it cannot be answered, such a question should not even be asked.

This, of course, is nonsense: such a question *can* be asked in this model, but the answer depends on the geometry. It is the Independence Fallacy which leads to a denial of an underlying reality—stemming from a motivation to describe a slice of a system independently from what lies outside.

Information-Based Updating

Leaving aside the Independence Fallacy, it should be clear how the $S_?$ description of the dotted box should be updated upon learning new information. For example, if an agent learned that the geometry was in fact that of Fig. 16.2b, a properly-updated description would simply be a 25/41 probability that the two coins were the same. And upon learning the actual values of the coins (say, *HT*), further updating would occur; *HT* would then have a 100 % probability.

But the central point is that some information-updating naturally occurs when one learns the geometry of the model, even without any revealed circles. And because this is a realistic model (with some real, underlying state), the information updating has no corresponding feature in the coin's objective reality. It is a subjective process, performed as some agent gains new information.

Introducing Time

The above model was presented as a static system in two spatial dimensions. The only place that time entered the analysis was in the updating process in the previous subsection, but this subjective updating had no relation to anything objective about the system. Indeed, one could give different agents information in a different logical order, leading to different updating. Both orders would be unrelated to any objective evolution of the system; after all, the system is static.

Still, an objective time coordinate can be introduced in a trivial manner: simply redefine the model such that one of the spatial axes in Fig. 16.2 represents time instead of space. Specifically, suppose that the vertical axis is time (past on the bottom, future on the top). It is crucial not to introduce dynamics along with time; one point of the model was to show how to analyze systems without dynamics. And since this analysis has already been performed, we don't need to do it again. The dotted box now represents an instantaneous slice, and the same state-counting logic will lead to exactly the same probabilities as the purely spatial case.

One might be tempted to propose reasons why this space-time model is fundamentally different from the original space-space model, perhaps assuming the existence of dynamical laws. Such laws *would* break the analogy, but they are not part of the model. Besides, the previous section is an existence proof that such a system *can* be analyzed in this manner, which is all that is needed for the below conclusions. It is logically possible to assign an equal probability to each temporally-extended microstate (or more intuitively, "microhistory") and then make associated predictions.

Sure, it's an open question whether there is some *other* way to analyze systems without dynamics, or if this approach has any chance of actually making *good* predictions. But this approach *is* empirically successful for spatial systems without dynamics, and the early indications are that it looks promising for temporal systems as well [10].

One unusual feature of the original model should now be obvious. Not knowing the spatial geometry (say, 2a or 2b) was an artificial restriction. But it's quite natural not to know the future, and once the vertical axis represents time, it's obvious why an agent might be uncertain whether the fourth circle would ever materialize. But this does not break the analogy between the spatial and temporal models. Sure, we tend to learn about things in temporal order, but it's not a formal requirement; we can film a movie of a system and watch it backwards, or even have spatial slices of a system delivered to us one at a time.[6] The link between information-order and temporal-order is merely typical, not a logical necessity.

In this temporal context, it's also more understandable how one might fall into the Independence Fallacy. If we expect the future to be generated from the past via some dynamical laws, then we would also expect the probabilities we assign to the past to be independent of the future experimental geometry. But without dynamics, if we assign every microhistory an equal probability, the standard information-updating that made sense in the spatial case also makes sense in the temporal case. When we learn about the experimental geometry of the future, this all-at-once analysis typically updates our probabilistic assessment of the past.

[6] As in the final section of [11].

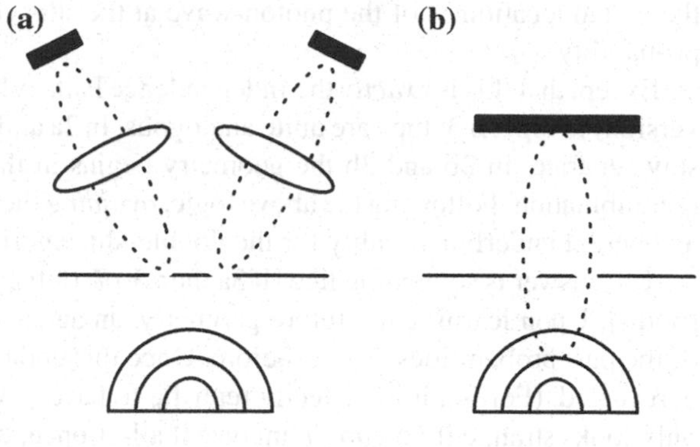

Fig. 16.3 Two geometries of a double slit experiment, in which a single photon passes through a pair of slits. **3a** Lenses and (*black*) detectors measure which slit the photon passes through; **3b** A screen records a photon that contributes to a two-slit interference pattern

Quantum Reality

Finally, we can apply this all-at-once analysis to quantum theory. This general approach has been suggested by various people over the years (see Appendix II for details of past research), but astoundingly, such proposals typically confound the matter by either giving up on A) ordinary spacetime, or B) ordinary probability theory. The below applications (and further arguments in [9]) demonstrate that these dramatic departures from classical physics are not necessary.

The modern arguments against an underlying reality for quantum systems typically involve hard-to-summarize "no-go theorems", but the central issues do not require anything so difficult, and indeed were well known to the founders of quantum mechanics. One example is the famous double-slit experiment. In Fig. 16.3, a source (at the bottom) creates a single photon that passes up through a pair of slits.[7] The classical concept most closely related to photons are classical electromagnetic waves/fields, but photons behave in a way that disagrees with the dynamical Maxwell equations which govern such fields. (A strike against dynamics.) Namely, photons always seem to be measured in particle-like chunks, rather than spread out as classically predicted. For example, when a lens (or two) images the slits (as in Fig. 16.3a), one always finds that the photon-wave went through one slit *or* the other.

And yet it appears that photons *do* spread out, at least between measurements, if one considers the experiment in Fig. 16.3b. Here a screen records the interference pattern produced by waves passing through *both* slits, built up one photon at a time. In the many-photon limit, this pattern is predicted by classical dynamics *only* if the waves pass through both slits and interfere. Since each individual photon conforms to this pattern (not landing in dark fringes), it seems evident that each photon also passes through both slits.

Where reality seems to fail here is the description of the photon at the slits—one slice of the full spacetime diagram. In 3a the photon seems to go through only one slit; in 3b it seems to go through both. And since the status of the photon at the slits is "obviously" independent of the future experimental geometry, it follows that

[7] The vertical axis is performing double-duty as both time and a second spatial axis.

the actual location(s) of the photon-wave at the slits cannot be assigned a coherent probability.

Except that this is *exactly* the Independence Fallacy! Compare Fig. 16.2 (temporal version) to Fig. 16.3; they are quite analogous. In 2a and 3a the right and left branches stay separate; in 2b and 3b the geometry begins in the same way, but then allows recombination. Following the above logic, *avoiding* the Independence Fallacy allows a coherent underlying reality for the double-slit experiment.

The answer is something like [If 3a then 50 % (left), 50 % (right); If 3b then 100 % (both)]. Upon learning the future geometry, an agent would update her assessment of the past probabilities, just as before. Once this updating occurs, a classical reality is revealed. (For 3b, it is perfectly realistic to have a wave go through both slits.) It only looks strange if you *don't* analyze it all-at-once, or attempt to map this process onto a story with dynamical evolution.

Unlike other resolutions of the double-slit experiment, *this* resolution naturally resolves more problematic situations. The no-go theorems against realistic models all use the Independence Fallacy in one form or another.[8] The typical assumption is that it's always fair to describe spatial slices independently from the future experimental geometry. But if one updates past probabilities upon learning which measurement a system will encounter, the premises behind these theorems are explicitly violated.

Even so, this complicated updating of probabilities on different time-slices is not the most natural picture. Relativity tells us that the slicing is subjective; the objective structure lies in the 4D spacetime block. It is here where the microhistories reside, and to be realistic, one of these microhistories must *really be there*. A physics experiment is then about *learning* which microhistory actually occurs, via information-based updating; we gain relevant information upon preparation, measurement setting, and measurement itself. And the best way to coherently describe this updating is with an all-at-once analysis.

Conclusions

If there is a plausible reality underlying quantum theory, the "It from Bit" idea looks wrongheaded. The microhistory-reality proposed here demands that one gives up the intuitive universe-as-computer story of dynamical time evolution, so one may still choose to cling to dynamics, voiding this analysis. But in the process, one is also rejecting a spacetime-based reality. Is this a fair trade-off? Is dynamics really so crucial that it's worth delving into some nebulous "informational immaterialism" [1] or elevating configuration space into some weird reality in its own right? And why should dynamical laws be so important if one is giving up on a fundamental reality in the first place?

After all, there are excellent reasons for dropping dynamics, the quantum no-go theorems being prime examples. We also have the beautiful path integral where *all*

[8] Outcome independence [12], preparation independence [5], etc.

possible histories must be considered (whether they obey dynamical laws or not; see, e.g., [13]). And is it really so crucial that we live in a universe where nothing interesting happens in the time-direction, where everything about the present was encoded in some initial cosmic wavefunction? It's not such a stretch to view our world as one *possibility* of infinitely many, unshackled from strict predeterministic rules.

After giving up on reality via the Independence Fallacy, the standard quantum story ironically responds by making almost everything *interdependent* in some strange configuration space. (Almost everything, just not the future or the past.) The simpler alternative proposed here is simply to link everything together in standard 4D spacetime. This casts our information in the classical form: $S = [p_1(W_1), p_2(W_2), \ldots, p_N(W_N)]$, with the crucial caveat that the W's are now micro *histories*, spanning 4D instead of 3D. So long as one does not additionally impose dynamical laws, there is no theorem that one of these microhistories cannot be real.

Still, qualitative arguments are one thing; the analogy between the above model and the double slit experiment can only be pushed so far. And one can go *too* far in the no-dynamics direction: considering *all* histories, as in the path integral, would lead to the conclusion that the future would be almost completely uncorrelated with the past, contradicting macroscopic observations.

But this approach can be made much more quantitative. The key is to only consider a large natural subset of possible histories,[9] such that classical dynamics is usually recovered as a general guideline in the many-particle limit. Better yet, for at least one model, the structure of quantum probabilities naturally emerges.[10] And as with any deeper-level theory that purports to explain higher-level behavior, intriguing new predictions are also indicated [10].

Even if the arguments presented in this essay are not a convincing reason to discard fundamental dynamical equations, they nevertheless serve as a strong rebuttal to the "It from Bit" proponents. Whether or not one *wants* to give up dynamics, the point is that one *can* give up dynamics, in which case quantum information can plausibly be *about something real*. Instead of winning the argument by default, then, "It from Bit" proponents now need to argue that it's *better* to give up reality. Everyone else need simply embrace entities that fill ordinary spacetime—no matter how you slice it.

Appendix I: The Universe is Not a Computer

Isaac Newton taught us some powerful and useful mathematics, dubbed it the "System of the World", and ever since we've assumed that the universe actually runs according to Newton's overall scheme. Even though the details have changed, we still basically hold that the universe is a computational mechanism that takes some initial state as an input and generates future states as an output.

---

[9] Those for which the total Lagrangian density is always zero.

[10] The Born rule can be derived for measurements on an arbitrary spin state in reasonable limits [10].

Such a view is so pervasive that only recently has anyone bothered to give it a name: Lee Smolin now calls this style of mathematics the "Newtonian Schema" [14]. Despite the classical-sounding title, this viewpoint is thought to encompass all of modern physics, including quantum theory. This assumption that we live in a Newtonian Schema Universe (NSU) is so strong that many physicists can't even articulate what other type of universe might be conceptually possible.

When examined critically, the NSU assumption is exactly the sort of anthropocentric argument that physicists usually shy away from. It is essentially the assumption that the way we solve physics problems must be the way the universe actually operates. In the Newtonian Schema, we first map our knowledge of the physical world onto some mathematical state, then use dynamical laws to transform that state into a new state, and finally map the resulting (computed) state back onto the physical world. This is useful mathematics, because it allows us to predict what we don't know (the future), from what we do know (the past). But it is possible we have erred by assuming the universe must operate as some corporeal image of our calculations.

The alternative to the NSU is well-developed and well-known: Lagrangian-based action principles. These are perhaps thought of as more a mathematical trick than as an alternative to dynamical equations, but the fact remains that all of classical physics can be recovered from action-extremization, and Lagrangian Quantum Field Theory is strongly based on these principles as well. This indicates an alternate way to do physics, without dynamical equations—deserving of the title "the Lagrangian Schema".

Like the Newtonian Schema, the Lagrangian Schema is a mathematical technique for solving physics problems. One sets up a (reversible) two-way map between physical events and mathematical parameters, partially constrains those parameters on some spacetime boundary *at both the beginning and the end*, and then uses a global rule to find the values of the unconstrained parameters and/or a transition amplitude. This analysis does not proceed via dynamical equations, but rather is enforced on entire regions of spacetime "all at once".

While it's a common claim that these two schemas are equivalent, different parameters are being constrained in the two approaches. Even if the Lagrangian Schema yields equivalent dynamics to the Newtonian Schema, the fact that one uses different inputs and outputs for the two schemas (i.e., the final boundary condition is an input to the Lagrangian Schema) implies they are not exactly equivalent. And conflating these two schemas simply because they often lead to the same result is missing the point: These are still two different ways to solve problems. When *new* problems come around, different schemas suggest different approaches. Tackling every new problem in an NSU (or assuming that there is always a Newtonian Schema equivalent to every possible theory) will therefore miss promising alternatives.

Given the difficulties in finding a realistic interpretation of quantum phenomena, it's perhaps worth considering another approach: looking to the Lagrangian Schema not as equivalent mathematics, but as a different framework that can be altered to generate physical theories not available to Newtonian Schema approaches [6].

Appendix II: Previous Work

In quantum foundations, analyzing four-dimensional histories "all at once" is uncommon but certainly not unheard of; several different research programs have pursued this approach. Still, in seemingly every one of these programs, the history-analysis is accompanied with a substantial modification to (A) ordinary spacetime, or (B) ordinary probability and logic. Looking at previous research, one might conclude that it is not the all-at-once analysis that resolves problems in quantum foundations, but instead one of these other dramatic modifications. But such a conclusion is incorrect; a history-based framework can naturally resolve all of the key problems without requiring any changes to (A) or (B).

Any approach that incorporates the standard quantum state is already a dramatic modification to spacetime (A), because multiparticle wavefunctions do not reside in ordinary spacetime. (They instead reside in a higher-dimensional configuration space.) This includes approaches that (arguably) have some all-at-once element (including GRW-style flash ontologies [15–17], Cramer's Transactional Interpretation [18] and the Aharonov-Vaidman two-state approach [19]). Even if these approaches somehow argued that they did not use the *standard* quantum state, they are still using functions on configuration space, not spacetime—and therefore fall in category (A).

Several history-based approaches in the literature do not take anything like the standard quantum state to be a "real" part of the theory. Griffiths' "Consistent Histories" framework [20] is one example, although it is not a full explanation, as there are many cases where no consistent history can be found. Also, there is never *one* fine-grained history that can be said to occur. Gell-Mann and Hartle have recently [21] attempted to resolve these problems, but in the process they modify probabilistic logic (B), enabling the use of negative probabilities.

Several history-based approaches have been championed by Sorkin and colleagues. A research program motivated and based upon the path integral [22] is of particular relevance, although it is almost always presented in the context of a non-classical logic (B). (Not all such work falls in this category; one notable exception is a recent preprint by Kent [23]). This path-integral analysis is rather separate from Sorkin's causal set program [24], which seeks to discretize spacetime in a Lorentz-covariant manner. While this is also history-based, it clearly modifies spacetime (A).

Another approach by Stuckey and Silberstein (the Relational Blockworld [25]) is strongly aligned against dynamical laws, and the all-at-once aspect is central to that program. But again, this history-based framework comes with a severe modification of spacetime (A), in that the Relational Blockworld replaces spacetime with a discrete substructure. It is therefore unclear to what extent this program resolves interpretational questions via the non-existence of ordinary spacetime rather than simply relying on the features of all-at-once analysis.

Finally, an interesting approach that maps the standard quantum formalism onto a more time-neutral framework is recent work by Leifer and Spekkens [26]. Notably, it explicitly allows updating one's description of the past upon learning about future events. Because the quantum conditional states defined in this work are clearly analogous to states of knowledge rather than states of reality, the fact that they exist in a large configuration space is not problematic (and indeed there is a strong connection to work built on spacetime-based entities [13]). Still, the logical rules required to extract probabilities from these states differ somewhat from classical probability theory (B).

Any of the above research programs may turn out to be on the right track; after all, there is no guarantee that the entities that make up our universe *do* exist in spacetime. But the fact that most of these approaches modify spacetime (or logic) has thoroughly obscured a crucial point: A history-based analysis, with no dynamical laws, *need not modify spacetime or logic to resolve quantum mysteries*, even taking the quantum no-go theorems into account. For further discussion of this point, see [9].

Appendix III: Model Details

The model in Fig. 16.2 (reproduced below) has the following rules. Each circle can be in the state heads (H) or tails (T), and each line connects two circles. Each line has one of three internal colors; red (R), green (G), or blue (B), but these colors are *unobservable*. The model's only "law" is that red lines must connect opposite-state circles ($H-T$ or $T-H$), while blue and green lines must connect similar-state circles ($H-H$ or $T-T$).

When analyzing the state-space, the key is to remember that connecting links between same-state circles have two possible internal colors (G or B), while links between opposite-state circles only have one possible color (R). Combined with the equal *a priori* probability of each complete microstate (both links and circles), this means that for an isolated two-circle system, the circles are twice as likely to be the same as they are to be different.

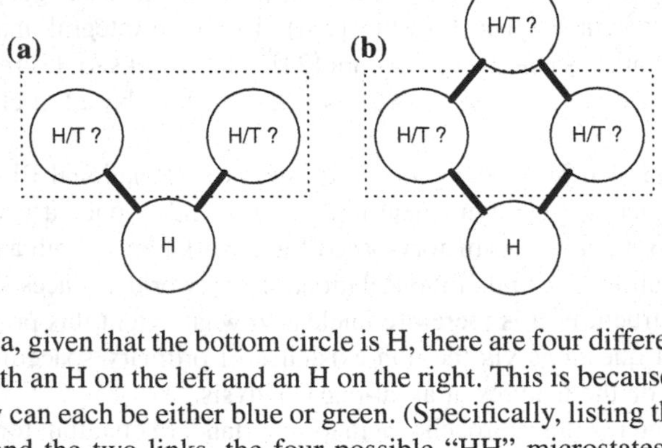

In Fig. 16.2a, given that the bottom circle is H, there are four different microstates compatible with an H on the left and an H on the right. This is because there are two links, and they can each be either blue or green. (Specifically, listing the states of the three circles and the two links, the four possible "HH" microstates are HBHBH, HBHGH, HGHBH, and HGHGH.) According to the fundamental postulate of statistical mechanics, an HH will be four times as likely as a TT, for which only red links are possible (TRHRT). The full table for Fig. 16.2a is:

Left	Right	Microstates
H	H	4
H	T	2
T	H	2
T	T	1
2a	Total:	9

Figure 16.2b is more complex, in that there is now a fourth circle at the top. The fact that there are four links also means that there are 16 different microstates corresponding to all H's (4 green or blue links, $2^4 = 16$), but only one microstate corresponding to the case with T's on the right and left and another H on the top (4 red links). The 2b table is:

Left	Right	Top	Microstates
H	H	H	16
H	H	T	4
T	H	H	4
T	H	T	4
H	T	H	4
H	T	T	4
T	T	H	1
T	T	T	4
	2b	Total:	41

However, since we are not interested in the status of the top circle in this model, the relevant numbers are the total number of ways in which one might have (say) an H on the left and right. To get the total number of such states, one simply sums the first two rows of the previous table. In other words, there are 20 different states that have HH in the dotted box of Fig. 16.2b; 16 with H on top and 4 with T on top. The more useful 2b table is therefore:

Left	Right	Microstates
H	H	20
H	T	8
T	H	8
T	T	5
2b	Total:	41

Notice there are 25 ways in which the right and left circles match, versus 16 ways in which they do not match. This contrasts with a 5:4 ratio for Fig. 16.2a.

References

1. C.G. Timpson, *Quantum Information Theory and the Foundations of Quantum Mechanics* (Oxford University Press, Oxford, 2013)
2. J.A. Wheeler, Information, physics, quantum: the search for links, in *Complexity, Entropy and the Physics of Information*, ed. by W.H. Zurek (Addison-Wesley, Redwood City, 1990)
3. J.S. Bell, On the Einstein Podolsky Rosen Paradox. Physics **1**, 195 (1964)
4. S. Kochen, E.P. Specker, The problem of hidden variables in quantum mechanics. J. Math. Mech. **17**, 59 (1967)
5. M.F. Pusey, J. Barrett, T. Rudolph, On the reality of the quantum state. Nat. Phys. **8**, 475 (2012)
6. K. Wharton, The universe is not a computer (2012). arXiv:1211.7081
7. D. Bohm, A suggested interpretation of the quantum theory in terms of hidden variables. Phys. Rev. **85**, 166 (1952)
8. H. Everett, Relative state formulation of quantum mechanics. Rev. Mod. Phys. **29**, 454 (1957)
9. K. Wharton, Quantum states as ordinary information. Information **5**, 190 (2014)
10. K.B. Wharton, Lagrangian-only quantum theory (2013). arXiv:1301.7012
11. P.W. Evans, H. Price, K.B. Wharton, New slant on the EPR-bell experiment. Brit. J. Found. Sci. **64**, 297 (2013)
12. A. Shimony, *Sixty-Two Years of Uncertainty: Historical, Philosophical, and Physical Inquiries into the Foundations of Quantum Mechanics* (Plenum, New York, 1990)
13. K.B. Wharton, D.J. Miller, H. Price, Action duality: a constructive principle for quantum foundations. Symmetry **3**, 524 (2011)
14. L. Smolin, The unique universe. Phys. World **22N6**, 21 (2009)
15. J.S. Bell, Toward an exact quantum mechanics, in *Themes in Contemporary Physics*, ed. by S. Deser, R.J. Finkelstein (World Scientific: Teaneck, New Jersey, 1989)
16. A. Kent, Quantum jumps and indistinguishability. Mod. Phys. Lett. A **4**, 1839 (1989)
17. R. Tumulka, A relativistic version of the Ghirardi-Rimini-Weber model. J. Stat. Phys. **125**, 821 (2006)
18. J.G. Cramer, Generalized absorber theory and the Einstein-Podolsky-Rosen Paradox. Phys. Rev. D **22**, 362 (1980)
19. Y. Aharonov, L. Vaidman, Complete description of a quantum system at a given time. J. Phys. A **24**, 2315 (1991)
20. R.B. Griffiths, Consistent histories and the interpretation of quantum mechanics. J. Stat. Phys. **36**, 219 (1984)
21. M. Gell-Mann, J.B. Hartle, Decoherent histories quantum mechanics with one 'Real' fine-grained history. Phys. Rev. A **85**, 062120 (2011)
22. R.D. Sorkin, Quantum dynamics without the wave function. J. Phys. A **40**, 3207 (2007)
23. A. Kent, Path integrals and reality (2013). Available online: arXiv:1305.6565
24. R.D. Sorkin, Scalar field theory on a causal set in histories form. J. Phys. **306**, 012017 (2011)
25. W.M. Stuckey, M. Silberstein, M. Cifone, Reconciling spacetime and the quantum: relational blockworld and the quantum liar paradox. Found. Phys. **38**, 348 (2008)
26. M.S. Leifer, R.W. Spekkens, Towards a formulation of quantum theory as a causally neutral theory of Bayesian inference. Phys. Rev. A **88**, 052130 (2013)

Chapter 17
Bit from It

Julian Barbour

Abstract With his aphorism 'it from bit', Wheeler argued that anything physical, any *it*, ultimately derives its very existence entirely from discrete detector-elicited information-theoretic answers to yes or no quantum binary choices: *bits*. In this spirit, many theorists now give ontological primacy to information. To test the idea, I identify three distinct kinds of information and find that *things*, not information, are primary. Examination of what Wheeler meant by 'it' and 'bit' then leads me to invert his aphorism: 'bit' derives from 'it'. I argue that this weakens but not necessarily destroys the argument that nature is fundamentally digital and continuity an illusion. There may also be implications for the interpretation of quantum mechanics and the nature of time, causality and the world (For publication in this volume, I have added some new footnotes, dated 2014, in which I indicate developments in my thinking since the essay competition, giving details of any appropriate publications. I also take the opportunity, omitted at the time, to respond to some of the comments that were made of my essay in FQXi posts (at http://fqxi.org/community/forum/topic/911). I have also, without noting them, made a few trivial changes to the text for the sake of greater clarity and precision).

Introduction

Quantum information theory has suggested to numerous researchers that the ground of being—ultimate reality—is information. John Wheeler [1] is the prophet of this movement. Vlatko Vedral [2] argues that "information is physical", and Paul Davies [3] suggests that information is 'real' and "occupies the ontological basement". Both argue that information is more basic than quantum fields or energy. Moreover, in line with Wheeler's 'it from bit', they take information, and with it reality, to be digital and to rest ultimately on the answers to yes/no questions. Continuity is an illusion.

To see if such proposals are likely to be correct, we need a definition of *information*. What is it? This is the first issue that I address. I distinguish three kinds of

J. Barbour (✉)
College Farm, The Town, South Newington, Banbury, OX 15 4JG, UK
e-mail: julian.barbour@physics.ox.ac.uk

A. Aguirre et al. (eds.), *It From Bit or Bit From It?*,
The Frontiers Collection, DOI 10.1007/978-3-319-12946-4_17

information: as defined by Shannon, as used in normal language, and as intrinsic semantic information. On this basis, I conclude that ontological primacy should not be given to information but to *things*, as has always been the standpoint of realists.

I also find it important to define 'bit' and 'it'. Wheeler's 'bit' is strictly something that belongs to our perceptions, while an 'it' is something like a quantum field or particle whose existence we deduce from a pattern of perceived bits. Consideration of everything involved in the deduction process, in which the nature of explanation plays an important role, makes me question Wheeler's contention that every 'it' derives *its very existence* from bits. I find no reason to reverse the standard assumption of physics, namely that what we experience can be explained by the assumption of an external world governed by law. On this basis, Wheeler's aphorism should be reversed: 'bit' derives from 'it'.

An important part of my argument relates to the nature of a thing, which I argue is necessarily holistic and must be complete. The definition of a thing then amounts to a description of the universe from a particular point of view.[1] At the end of my essay, I consider how this bears on the interpretation of quantum mechanics and the nature of time and the world.

General Comments

A symbol can stand for *anything*, but it must stand for *something*. Thus, x can stand for the position of a particle, and the digit 1 can stand for one apple, one pear, etc. Otherwise 1 is just black ink on white paper—and that too is something. This is important because Wheeler argues that "rocks, life, and all we call existence" are based on immaterial yes/no bits of information. It is a mistake to believe that the digits 0 and 1, being abstract, represent the immaterial. Quite to the contrary, I shall show that they stand for something quintessentially concrete.

My arguments rely on definite meanings of 'real' and 'existing'. Here it is necessary to distinguish what we experience directly from things that we hypothesize to explain what we experience. I know my conscious experiences are real and exist because I have direct access to them. Bishop Berkeley's maxim "To be is to be perceived" applies. Berkeley argued that, even as scientists, we do not need to postulate a real world behind experiences. This is the philosophy of idealism. According to it, the proper task of science is merely to establish the correlations between experiences. However, the success of theory in science suggests rather strongly that assuming the existence of things that we cannot see to explain things that we can is a good

[1] 2014: The 'from a particular point of view' is unfortunate and in conflict with my actual position expressed in Sect. General Comments. It slipped in through my enthusiasm for Leibniz, for whom sentient beings (monads) are the 'true atoms of existence'. As I will spell out in further footnotes, especially 14, below (which take into account quantum mechanics), my working assumption is now rather precisely that our internal perceptions are a partial reflection—a particular point of view—of an external universe that is a collection of shapes. Both the individual shapes and their collection, which I call *shape space*, are holistic concepts.

strategy. A striking early example comes from ancient astronomy. Greek astronomers observed intricate motions of the sun, moon, and planets on the *two*-dimensional sky. They explained them—saved the appearances—by positing simple regular motions of the celestial bodies in *three* dimensions. The success of the enterprise, brought to a triumphant conclusion by Kepler [4], justified belief in the reality of the assumed motions and extra dimension. Many more examples like this could be given. The findings of science are always provisional, but my position is that something that one does not directly observe exists if it explains phenomena.[2]

Kinds of Information

It is perhaps a pity that Shannon's *The Mathematical Theory of Communication* [5] somehow morphed into *information theory*. That gives rise to potential confusion between three different meanings of 'information'. Let us distinguish them carefully.

The first is *Shannon information*, which he also called entropy or uncertainty. It involves *things* (which Shannon called messages) and *probabilities* for those things. Both are represented abstractly, the things by symbols (in practice binary numbers) and the probabilities by numbers. In line with what I said about abstraction, the symbols have no meaning if divorced from the entities that they represent. The same is true of the probablities; we need to know their method of determination.

This is well illustrated by the antecedant of Shannon's theory: the code that Morse developed in the 1830s. To increase transmission speed, Morse chose the length of his symbols for the various letters of the alphabet broadly according to the frequency (probability) with which they occur. Unlike Shannon, who did a thorough statistical analysis of English text, Morse simply estimated the relative frequencies of letters "by counting the number of types in the various compartments of a printer's type box" [6]. This shows that the Morse code was born as an abstract amalgam of two elements: pieces of lead type and the relative numbers of them in a printer's box.

In close analogy, Shannon's theory of communication considers *a source of messages* and *the probabilities with which they are chosen*. The universality of the theory rests on the multitude of things that can serve as messages: the letters of any language (with their obvious antecendant in the printer's box), the words of that language, sentences in that language, and continuous or discrete distributions in, say, a two-dimensional field of view. All such messages correspond to concrete things:

[2] 2014. In a discussion post, Tom McFarlane commented "there are often multiple distinct explanations for the same phenomena, i.e., under-determination of theory by the facts". In response, I would distinguish between contrived theories, which one need not take seriously, and more solid theories. For example, many theoreticians are currently exploring sensible alternatives to Einstein's general theory of relativity that are all compatible with presently known phenomena. To the extent they succeed, one can have *provisional* belief in the external world they assume. My general thesis does not depend on the specific kind of external reality in which currently I personally believe.

type is structured lead, words stand for things,[3] and sentences for concatenations of words. Distributions are particularly important in the context of this essay. First, they establish a direct correspondence with our most immediate experiences. For every time we open our eyes, we see a distribution of coloured shapes. Such a distribution is one of the messages that nature is constantly communicating, Shannon-like, to our consciousness. Second, perceived distributions directly suggest the most fundamental ontological concept in theoretical physics: a field configuration, which in the simplest example of a scalar field can be likened to a field of variable light intensity.

On the face of it, the probabilities that, as we shall see, form such an important part in Shannon's theory, are very different from the messages, which stand for things. However, the probabilities on which Shannon based his theory (and are relevant for this essay) were all based on objective counting of relative frequencies of definite outcomes. The example of Morse proves that: he counted the numbers of different type in the printer's box. So the probabilities too have an origin in things: although they do not stand for things, they stand for proportions of things in a given environment. This is just as true of observationally determined quantum probabilities as it is for the frequencies of words in typical English or the numbers of different trees in a forest.

The concepts of message and probability enable one, for a definite source of N messages, to define Shannon's *information*.[4] If p_i, $i = 1, 2, \dots, N$, is the relative probability of message i and $\log p_i$ is its base-2 logarithm, then the information I of the given source is

$$I = -\sum_{i}^{N} p_i \log p_i. \tag{17.1}$$

The minus sign makes I positive because all probabilities, which are necessarily greater than or equal zero, are less than unity (their sum being $\sum_i^N p_i = 1$), so that their logarithms are all negative (or zero in the case of certainty).

The definition (17.1) is uniquely fixed by a few desirable or essential properties such as positivity and additivity; the most important is that I takes its maximum value when all the p_i are equal; the slightest deviation decreases I. If an unbiased coin, with equal probabilities $p_h = p_t = 1/2$ for heads and tails, is tossed once, the value of I is unity and defined as one *bit* of information (or uncertainty). The number of possible outcomes (messages) increases exponentially with the number N of tosses: The information I is then N and equal to the base-2 logarithm of the number M of different outcomes. For five tosses $M = 2 \times 2 \times 2 \times 2 \times 2 = 32$ and $\log_2 32 = 5$. If a coin is biased, $p_h \neq p_t$, then for a single toss I is less than one, becoming zero when only one outcome is possible. All this is rather simple and beautiful—once Shannon had the idea.

[3] Verbs by themselves have no meaning. In the sentence "Bit dog man" (the standard order in Irish Gaelic), we would not know what 'bit' means had we not seen canine teeth in action.

[4] On von Neumann's advice, Shannon also called it *entropy* by analogy with Boltzmann's entropy in statistical mechanics.

Pierce [6] carefully distinguishes the *binary digits* 0 and 1 from Shannon's information bit. There are two good reasons to do this. First, the digits 0 and 1 can serve as messages, but their probabilities of transmission may be far from equal. The information of the source will then be less than one. Second, the name binary digit for either 0 or 1 may, as John Tukey suggested, be contracted to 'bit', but one needs *equal* uncertainty associated with *two* things (which may be the two binary digits) to get *one* information bit. Pierce's distinction is useful too in helping to clarify what 'it' and 'bit' mean. Wheeler is explicit: bits are detector-elicited answers to yes or no quantum binary choices. Now an answer in quantum experiments is essentially a binary digit as defined by Pierce: for example, 0 will stand for spin down and 1 for spin up. This would be the case even if the two possible outcomes do not have equal probabilities.

The information-theoretic (17.1), *Shannon information*, is quite different from what most people mean by information and I call *factual information*. This last is actually the *content* of Shannon's messages. It can be anything: a string of random binary digits, instructions to a bank, or the news that President Kennedy has been assassinated. If we receive a picture, we normally understand by information the distribution of colours and shapes we see when looking at it. I have already likened a distribution to a *configuration*, but the word can stand for any *structured* thing. Structure and variety are central to my critique of 'it from bit'. For we can only talk meaningfully about a thing, including a 'bit', if it has distinguishing attributes. The way that they are knit together, as in the taste, shape and colour of an apple, defines the structure of the thing. There is one metalaw of science: it cannot exist without structured things. Structured variety is the ground of being. That is what gives content to both science and life.

Having identified configurations as examples of Shannon messages that carry factual information, I now come to the *intrinsic semantic information* they may have. Consider the example of a *time capsule*, which I discuss in [7]. I do not mean "a container holding historical records . . . deposited . . . for preservation until discovery by some future age" (Webster's) but something that arises naturally. My favourite example comes from geology.[5] Two centuries ago, geologists started to establish detailed connections (near congruences in the first place) between the structure of fossils and rocks found in different locations. They concluded that the connections could only be explained by an immense age of the earth, vastly longer than the bible-deduced estimate of somewhat over 6000 years. They explained the earth's present state by a long process that had unfolded in accordance with the then known laws of nature. They discovered deep time. The geologists' discovery of physical history and the present evidence for it extends today most impressively to all branches of science, especially cosmology and genetics.

Everywhere scientists look they find records that speak with remarkable consistency of a universe which began in a very special big bang—with the geometry of space highly uniform but the matter in higgledly-piggledy thermal equilibrium—and has since evolved, creating in the process a record of what happened. For the sake

[5] Appendix A gives a different example of semantic content generated by a timeless law.

of a concept to be used later, suppose one instantaneous configuration of the universe is recorded at the present cosmological epoch. Its specification (without any momentum information) is consistent with quantum mechanics and relativity.[6] In a coarse-grained form it will contain all the factual information that the geologists used to make their discovery. In fact, thanks to the stability of solids, the fossils and rocks that they used exist to this day essentially unchanged. I claim that the configuration carries intrinsic semantic information in the sense that different intelligent beings can in principle deduce *the law or process that explains the observed structure*. Support for this is the independent discovery of evolution by natural selection by Wallis and Darwin through their common reaction to the same evidence in fossils and living animals.

In summary, we must distinguish three kinds of information: Shannon's information, the uncertainty as to which message will be selected from a source; factual information, the content of such a message; and intrinsic semantic information, which distinguishes a random message, or configuration, from one that carries meaning and to some extent explains its very genesis. *All three have a firm underpinning in things.* Finally, Shannon-type message sources could not exist if the universe were not subject to laws of nature and far from thermal equilibrium.

This applies in particular to the quantum 'incarnation' of a Shannon one-bit information source: the *qubit* (quantum bit). Qubits are perfect information-theoretic bits, but they do not 'float around' in the universe ready to be put to use as message sources, any more than unbiased coins could be found in nature before humans invented them as value tokens. Qubits can be realized in many different ways [8], but all require great experimental sophistication and rely on the low entropy of the universe.

The Status of Information

Having defined the three kinds of information, I can draw my first conclusion: information, in Shannon's sense, must have an underpinning in things. Information theory would never have got off the ground if structured things—configurations—did not exist. In a forest we can count trees and establish their relative numbers. This is just what Morse and Shannon did with type and letters. Probabilities without things are pure *no*things. It is also relevant that the *outcomes* of the experiments used to establish quantum probabilities are determined in the macroscopic classical world in essentially the same way as the outcomes of classical processes.[7]

The key point is this. If we are to speak about ontology, as opposed to efficient coding in communication channels, the most important symbol in (17.1) is not p for

[6] In relativity, fundamentally defined 'nows' are denied. However, in the actual universe cosmologists can and do define them using the distribution of matter.

[7] In the absence of a viable hidden-variables explanation of quantum phenomena, it does appear that quantum probabilities have a better 'birthright' than the ignorance of classical physics to appear in the expression (17.1) for information. However, I do not think that this affects the main thrust of my argument.

probability but i for the thing, or configuration, that has the probability p_i. Probabilities are for outcomes: what you find when you open the box. Thus, even if quantum probabilities are an integral and essential part of the world, they are meaningless in themselves. They are at best secondary essentials, not primary essentials. They must always be probabilities for *something*.

Now what does this tell us about the world and how we should conceptualize it? It is clear that quantum mechanics must be taken into account. If an experimentalist has prepared a simple two-state quantum system, a qubit, it can serve as a Shannon-type information source. The two possible outcomes are not yet factual, and at this stage the state is an inseparable amalgam of two things: outcomes and the probabilities for them. If information is understood in this strict Shannon sense, Zeilinger is right to say, in his beautiful description of teleportation experiments [9], that "the quantum state that is being teleported is nothing other than information". He then goes on to say, as an observation which he regards as very important, that "the concepts *reality* and *information* cannot be separated from each other." However, in his book Zeilinger does not define information or distinguish the kinds of information that I have so far described; by 'information' he generally seems to mean factual as opposed to Shannon information (though not in the teleportation example). I have found that other authors do make the distinction but often fail to maintain it, so that one is left trying to make out which kind of information they mean.

But whatever authors may mean by information, quantum states still give us probabilities for outcomes in the form of factual information about things. Moreover, the probabilites themselves are determined by observation of things. I therefore conclude that things are the ground of being and constitute the ontological basement. Reality creates information and *is* separate from it. Once this has been recognized, we see that, for all its importance, information theory can in no way change what has always been the starting point of science: that structured things exist, in the first place in our mind and, as a reasonable conjecture given the remarkable correlations in our mental experiences, in an external world. Moreover, the proper task of ontology is to establish *the structure of things*. To this we now turn. It brings us back to quantum yes/no experiments and the need to establish what Wheeler meant by 'it' and 'bit'.

Wheeler's It and Bit

Wheeler explains what he means by 'it from bit' in, for example [1]. He describes an Aharonov–Bohm experiment in which an electron beam passes either side of a coil in which a magnetic flux is confined. Behind the coil is a screen on which individual electron hits are registered. The hits are found to form the characteristic fringes that arise from the quantum interference between the electron waves that pass on the two sides of the coil. When no flux is present in the coil, the fringes occupy a certain position. When a flux is present, the fringes are displaced, and from the magnitude of the displacement one can measure the strength of the magnetic field in the coil. By 'it' in the present case, Wheeler means the magnetic field and claims that it "derives its very existence entirely from discrete detector-elicited and information-theoretic answers to yes or no quantum binary choices: *bits*".

This needs some unpacking, above all of the bits. They are clearly the individual detector-elicited electron hits. The setup forces each electron to make a quantum choice as to where it will be registered, and it is from the totality of the hits that the field strength is deduced. We see that Wheeler's 'bit' is factual information. It gives the location of an electron impact. It is not a Shannon bit. Wheeler writes as if his 'bit' were one of the *binary digits*, 0 or 1, in the sense that Pierce defined them. But, in fact, it is not a single digit as could be used to record a genuine yes/no quantum outcome, say spin up or down. In this (typical) example, the record of the outcome is not one digit but a binary number of potentially infinite length because it needs to define the coordinates of the hit. To turn the hit into a yes/no (Boolean) answer, one must ask: did the electron impact within a specified area on the screen? But now infinitely many digits come into play *implicitly* through the definition of the area.[8]

At this point, we need to ponder the very conditions that make science possible: our special place in a special universe. Above all, the universe has very low entropy. The huge range of temperatures found in different locations is sufficient evidence of that. This is why on the earth's surface we can take advantage of the solid state to perform experiments and record their outcomes, either as dots on a screen or in a computer memory. Science would be impossible without the solid state. It would also be impossible without local inertial frames of reference, in which Newton's first law holds to a good accuracy. It keeps equipment in place. The low entropy of the universe and the law of inertia are fundamental properties of the universe and both can only be understood holistically (Appendix B).

The relation of what I have just said to Wheeler's 'bits' is this. They do not exist in isolation. A 'bit' is not a single-digit 'atom of reality' as 'it from bit' implies. A dot on a screen is not the unadorned answer to a straight question. A 'bit' has no meaning except in the context of the universe.

Consider Wheeler's Aharonov–Bohm experiment. When the flux in the coil is deduced, many things must be present, including knowledge of the laws of physics, physicists to use it, and the most directly relevant information: the pattern of the interference fringes with and without flux. It is only from the difference that a conclusion can be drawn. But we also need stability of the equipment, the laboratory and its environment, which must be monitored. All this requires conditions under which records can accumulate—in brief the conditions under which time capsules can form.

Wheeler's thesis mistakes abstraction for reality. Try eating a 1 that stands for an apple. A 'bit' is merely part of the huge interconnected phenomenological world that

[8] 2014: In a thoughtful post on my comments here, Christinel Stoica defends Wheeler and argues that his It-from-Bit idea must be understood in the conceptual framework of 'delayed-choice' experiments. Christinel's observations certainly help to clarify Wheeler's intuition (and I recommend the reader to have a look at them), but I feel my critique of Wheeler's aphorism stands: his hits of electrons on screens are not Shannon bits and they are not individual binary digits either. It might have helped had I been more explicit about my basic 'many-instants' interpretation [7] of quantum mechanics.

we call the universe and interpret by science; it has no meaning separated from that complex. Just because the overall conditions of the universe enable us to observe them in carefully prepared experiments, dots on screens are no proof that at root the world consists of immaterial single-digit information. For we have no evidence that the dots could exist in the absence of the world and its special properties.

The status of Wheeler's 'it' is not controversial and requires less discussion. It is the underlying invisible world, the world of quantum fields and particles, whose existence we deduce from the correlations within the interconnected phenomenological world. Of course, quantum mechanics indicates that, in any given measurement, we can only determine half of the classical variables associated with these entities, say the position or momentum of a particle but not both at once. However, the notions of particle and field remain crucial to our interpretation of quantum phenomena. Without them, we would understand nothing. In the quantum mechanics of a system of particles the wave function is defined on the space of possible configurations of the particles. The probabilities are then determined by the configuration space and the Schrödinger wave equation. Just as in classical mechanics, laws and configurations retain their indispensable explanatory role. They are the 'its' that explain phenomena.

Crucially, even if individual quantum outcomes are unpredictable, the probabilities for them are beautifully determined by a theory based on 'its'. I noted that a Shannon bit is an amalgam of things and probabilities and that the things have the deeper ontological status. Quantum mechanics strengthens this claim: the things determine the probabilities.[9] I see nothing in Wheeler's arguments to suggest that we should reverse the mode of explanation that has so far served science so well.

This conclusion in no way settles whether nature is analog or digital; that would at the least require us to know if the 'its' are continuous or discrete.[10] However, I do think that my arguments undermine some of the more extreme forms of 'digitalism'. Abstraction creates the impression that the world is made of qubits, but humans make qubits, just as they make coins.

Holism and Reductionism

What are the implications of this for quantum mechanics and the nature of time and the world? Some time before he published the *Principia*, Newton wrote a paper entitled "The Laws of Motion. How solitary bodys are moved" [4]. The 'solitary' indicated that Newton would define motion relative to invisible space and time. His resulting law of inertia became the prop of reductionism; it suggests that the most essential property of a body can be established by abstracting away everything in the universe that is observable. The catch, all too often forgotten, is that an inertial frame

[9] For the arguments of the following section, it is important that qubits are in stationary quantum states, for which the probabilities are defined by the *time-independent* Schrödinger equation. One has *timeless* probabilities for *possible* configurations.

[10] Nature may be both; space appears to have *three* dimensions but to be continuous.

of reference is needed to define the motion. If one asks after its origin, one is led to an account of motion in which configurations, not bland empty space, determine local inertial motion (Appendix B). I believe this undermines reductionism.

It also calls for a definition of *the universe*. I define it [7] as a set of *possible configurations* that nature has selected for reasons, perhaps of simplicity and consistency, that we have not yet fathomed.[11,12] The configurations are possible *instants of time*, In this picture, there is neither a containing space nor an unfolding time. One cannot ask when or where things happen; things are their own time and place. The Greek word *onta*, from which 'ontology' derives, means both 'existing things' and 'the present'.

Such a picture may help us to understand quantum mechanics. Zeilinger [9] emphasizes that the individual quantum measurement is purely random. He comments: "This is probably the most fascinating consequence in quantum physics . . . centuries of the search for causes . . . lead us to a final wall. Suddenly, there is something, namely the individual quantum event, that we can no longer explain in detail." But perhaps notions of causality have too long been tied to the picture of a world evolving in space and time. Leibniz argued long ago that the world is rational and that things are judged on their merits. The criterion for existence is comparison of 'possibles', not what was set up in some conjectured past. Leibniz also argued that reality resides in configurations, not position of bodies in space [16].

Newton said of himself "I do not know what I may appear to the world, but to myself I seem to have been only a boy playing on the seashore, and diverting myself in now and then finding a smoother pebble or prettier shell than ordinary, whilst the great ocean of truth lay all undiscovered before me." Is nature like Newton? There are many pretty shells among the possible configurations. What we observe and interpret as the outcome of an individual quantum event does not reside in space and time; it is embedded in a configuration. Nature's concern will surely be the big picture. In

[11] As of now, the best candidates are configurations of fields defined on a manifold that carries a three-metric. In essence, these are not unlike what we see when we open eyes.

[12] 2014: At the time this essay was being written, I did not refer to an important change in my thinking about time and configurations that was taking place. This was partly because my ideas were still developing but more importantly because the change did not in any way affect the main argument of the essay. However, let me here briefly indicate the nature of the change, which can be illustrated by the problem of three point particles that interact through Newton's law of gravitation. At any instant, the three particles form a triangle, which can be represented by two angles, which define its *shape*, and a further number which defines its size (area). Now the latter is a dimensionful quantity that depends on an arbitrary definition of the unit of length, say inch or centimeter. If the three particles are taken to model the universe, size can have no meaning. One needs to make a clear distinction between the shape of the triangle, defined by two dimensionless angles, and its size. By configuration in my earlier work, I included the size of the considered system. In my more recent work with collaborators, I have come to regard only the shape as fundamental. We formulate the dynamics of the universe as *shape dynamics*. Size still plays a role, but only as a dimensionless ratio of sizes at different epochs. This ratio then appears as an internal time variable. I cannot go into the many interesting consequences of this change of perspective, but instead refer the reader to the papers [10–15]. I will only say that if I were to write the essay now, I would replace the word 'configuration' in what follows by 'shape'. With one exception that I will indicate, the points I want to make would not be changed.

the timeless set of *onta*, the possible configurations of the world, nature will not find the pretty shells 'now and then' but 'here and there'. And causality through time will be replaced by selection according to timeless probabilities.[13] That is still causality but of a different kind.

I have a Leibnizian conjecture [7] that might even have pleased Newton. It is this. The set of all *onta* is the ultimate Shannon source. The semantic content of the *onta* measures their probabilities.[14] Consciousness is the communication channel from the *onta* to our experiences.

My thanks to David Deutsch for a helpful discussion, to Roberto Alamino, Harvey Brown and Nancy McGough for comments, and to Boris Barbour for the figures.

---

[13] 2014: In line with the comments made in footnote 12, I would now express this somewhat differently since the presence of an internal time is compatible with a certain notion of causality. Many pages would be needed to make this clear. For the moment I refer the reader to [15].

[14] This idea requires amplification! Let me just say that the time-independent Schrödinger equation determines *the most probable shapes* of molecules. It is at least possible [7] that an analogous timeless equation for the universe gives the highest probability to configurations with shapes that carry semantic information. 2014: I think this could still be true with a time-dependent, as opposed to timeless, equation of the kind we expect in shape dynamics (see footnote 12 and [15]), though now it will be necessary to say that nature will find pretty shells both 'now and then' as well as 'here and there'. This gives me an opportunity to keep to the promise made in footnote 1 and also respond to the post of Anonymous, who doubted whether "all information exists in external reality and is merely transmitted to the consciousness ...Some information is internally generated." I'm not sure where Anonmymous draws the line between 'external' and 'internal', but for me the latter is everything that I experience. I am acutely aware that, at any instant, I have no control over what I will actually experience in the next instant though it does seem to fit into an at least partially coherent history. I assume that there is *psychophysical parallelism*, meaning that to everything which I experience there corresponds a mathematically structured external reality in which what I experience is 'encoded'. Thus, a 'super-mathematician' who could examine the full external reality could deduce what I experience in any instant. The external reality I conjecture is *dualistic*: a timeless realm of possible shapes of the universe and a wave function ψ defined on it that passes through a succession of states. By this I mean that in a given state each shape has a certain value of ψ and that the distribution of the $|\psi|$ values over the possible shapes changes continuously from one state to the next. In [7] I suggested that the values of ψ are defined on configurations and are timeless: fixed once and for all. I further conjectured that the distribution of the ψ values is very nonuniform, with the Born probability amplitude $|\psi|^2$ being concentrated on what I called *time capsules*: configurations whose structure is so special as to suggest that they must have arisen through a history governed by definite physical laws. I assumed further that our brains are, in any instant, embedded, in such time capsules and are themselves time capsules. Then my conjectured psychophysical parallelism related our vibrant experience of change to the rich structure of our time-capsule brain. To be precise, I conjectured that consciousness transforms information in static structure—being—into experienced change—becoming. I am still trying to digest the implications of my new belief that there is an effective internal time in the physics of the universe. One thing at least I can say is that the results of [15] do suggest that the wave function of the universe may well have an intrinsic propensity to give high values of $|\psi|^2$ to shapes of the universe that are time capsules.

Appendix A: Maximal Variety

In the body of the essay, I describe 'time capsules' as configurations that carry intrinsic semantic content. Here I give another example. It does not suggest history, but the notion of a structured configuration is all important.

Leibniz told princesses that we live in the best of all possible worlds, but in his serious philosophy he argued that we lived in the one that is *more varied* than any possible alternative (*Monadology*, §58). Some years ago, Lee Smolin and I found a way to express this idea in a simple mathematical model. One representation is in terms of a ring of N slots (here $N = 24$) into which balls of different colours (or darkness in grey shades) can be placed (Fig. 17.1).

The diagram is maximally varied in the sense that, without pointing to a particular ball or naming colours or sense of direction, each ball is more readily identified than for any other occupation of the slots. The rule of creation is minimization of indifference (the inverse of variety). Balls are identified by their neighbourhoods: the seven-slot neighbourhood centered on ball x at noon is either the string $SDDxDSD$ or $DSDxDDS$ (left-right symmetry), where S and D are same (S) or different (D) neighbours (colour symmetry). The indifference I_{ij} of slots i and j is equal to the length (3, 5, 7,...) of the respective strings needed to distinguish them. The total indifference of a distribution is $I = \sum_{i<j} I_{ij}$. The most varied distribution is the one for which I has its smallest value. The relative number of balls of each colour is not fixed in advance but found by the minimization of I. Figure 17.1 is typical of the maximal-variety configurations (in general there are 2 or 3 for each N, though for 24 Fig. 17.1 is the unique configuration). Interestingly, the symmetric rule of creation

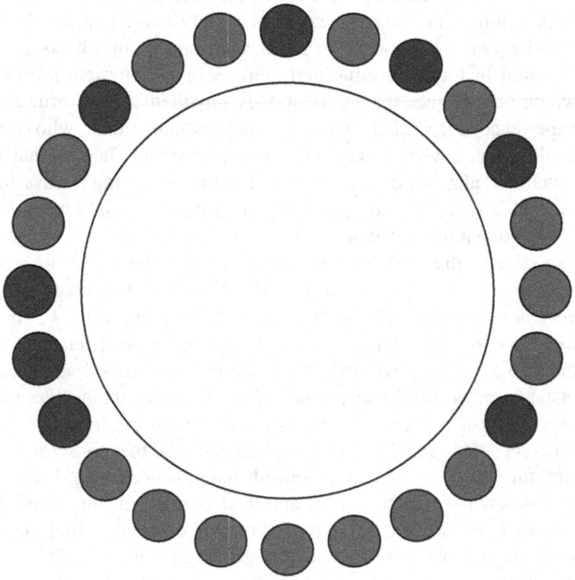

Fig. 17.1 The most varied two-colour (*dark-light*) 24-slot universe

invariably leads to markedly asymmetric configurations. The maximally—and near-maximally—varied configurations 'proclaim their own sematics' in the sense that smart enough mathematicians could deduce the law that creates them. For more details and the results of calculations up to $N = 27$ see [17].

Appendix B: Best Matching

This appendix describes the motion of particles without the props of absolute space and time. Position and time are treated relationally. The basic idea—best matching—also leads to general relativity (GR) [18]. The further requiremnt of relativity of size leads to GR with a distinguished definition of simultaneity [11]. I relied on this result in Sect. Holism and Reductionism to argue that identifying instants of time with confgurations is not in conflict with Einstein's theory.

Consider N gravitating particles (masses m_i) in Euclidean space. For $N = 3$, Fig. 17.2 shows the three particles at the vertices of the triangles representing two possible relative configurations 'held in imagination' somehow relative to each other. This generates apparent displacements $\delta\mathbf{x}_i$ of particle i. Calculate (17.2):

$$\delta A_{trial} = 2\sqrt{(E - V)\sum_i \frac{m_i}{2}\delta\mathbf{x}_i \cdot \delta\mathbf{x}_i}, \; V = -\sum_{i<j}\frac{m_i m_j}{r_{ij}}. \tag{17.2}$$

Here E is a constant. The action (17.2) is clearly arbitrary with no significance, but, using Euclidean translations and rotations, we can move either triangle relative to the other into the unique *best-matched* position that minimizes (17.2). For any two nearly identical triangles, the best-matched value of (17.2) defines a 'distance' between them and a metric on $\mathcal{S}$, the space of all possible relative configurations of the particles. One can then find the geodesics in $\mathcal{S}$ with respect to this intrinsically—and holistically—defined metric of the N-body universe.

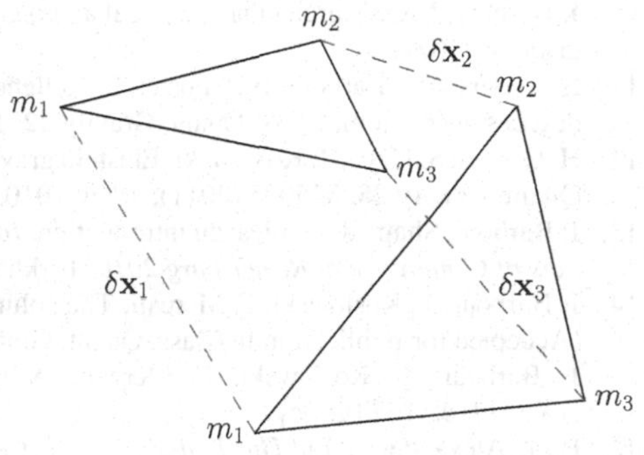

Fig. 17.2 A trial placing of the two *triangles* generates apparent displacements $\delta\mathbf{x}_i$. Minimization of the *trial action* (17.2) leads to the best-matched displacements

It can be shown [17] that the relative motions which the particles undergo as the representative point of the system moves along the geodesic are identical to the observable relative motions of particles in the Newtonian N-body problem with total energy E and *vanishing total angular momentum*. Moreover, the increment δt of 'Newtonian' time is *derived* [19]:

$$\delta t = \sqrt{\frac{\sum_i m_i \delta \mathbf{x}_i \cdot \delta \mathbf{x}_i}{2(E - V)}}. \tag{17.3}$$

The combination of best matching and the geodesic requirement, which dispenses with Newton's absolute time as independent variable, leads to recovery of Newtonian behaviour of subsystems in a large N-body 'island universe'. Newton's first law, the basis of reductionism, is holistic in origin.

References

1. J.A. Wheeler, Sakharov revisited: "It from Bit". *Proceedings of the First International A D Sakharov Memorial Conference on Physics, Moscow, USSR*, ed. by M. Man'ko (Nova Science Publishers, Commack, 1991), pp. 27–31
2. V. Vedral, *Decoding Reality: The Universe as Quantum Information* (Oxford University Press, Oxford, 2010)
3. P.C. Davies, N.H. Gregerson (eds.), *Information and the Nature of Reality* (Cambridge University Press, Cambridge, 2010)
4. J. Barbour, *The Discovery of Dynamics* (Oxford University Press, Oxford, 2001)
5. E. Shannon, W. Weaver, *The Mathematical Theory of Communication* (University of Illinois Press, Urbana and Chicago, 1963)
6. J.R. Pierce, *An Introduction to Information Theory*, 2nd edn. (Dover Publications, New York, 1980)
7. J. Barbour, *The End of Time* (Weidenfeld and Nicolson, London, Oxford University Press, New York, 1999)
8. D. Deutsch, It from qubit, in *Science and Ultimate Reality*, ed. by J. Barrow, D. Davies, C. Harper (Cambridge University Press, Cambridge, 2003)
9. A. Zeilinger, *Dance of the Photons* (Farrar, Strauss and Giroux, New York, 2010)
10. J. Barbour, N.Ó. Murchadha, Classical and quantum gravity on conformal superspace (1999), gr-qc/9911071
11. E. Anderson, J. Barbour, B.Z. Foster, B. Kelleher, N.Ó. Murchadha, The physical gravitational degrees of freedom. Class. Quant. Gravity **22**, 1795–1802 (2005), arXiv:gr-qc/0407104
12. H. Gomes, S. Gryb, T. Koslowski, Einstein gravity as a 3D conformally invariant theory. Class. Quant. Gravity **28**, 045005 (2011), arXiv:1010.2481
13. J. Barbour, Shape dynamics: an introduction. *In Quantum Field Theory and Gravity. Proceedings of Conference at Regensburg 2010* (Birkhäuser, 2012), ed. by F. Finster et al.
14. J. Barbour, T. Koslowski, F. Mercati, The solution to the problem of time in shape dynamics (Accepted for publication in Class. Quant. Gravity 2014) (2013), arXiv: 1302.6264 [gr-qc]
15. J. Barbour, T. Koslowski, F. Mercati, A gravitational origin of the arrows of time. arXiv:1310.5167 [gr-qc]
16. H.G. Alexander (ed.), *The Leibniz-Clarke Correspondence* (Manchester University Press, Manchester, 1956)

17. J. Barbour, The deep and suggestive principles of Leibnizian philosophy (downloadable from platonia.com). Harvard Rev. Philos. **11**, 45–58 (2003)
18. J. Barbour, Dynamics of pure shape, relativity and the problem of time. Lect. Notes Phys. **633** 15–35 (2003), gr-qc/0309089
19. J. Barbour, The nature of time, Winner of first juried prize essay of The Foundational Questions Institute (fqxi.org) (2009), arXiv:0903.3489 [gr-qc]

Chapter 18
Contextuality: Wheeler's Universal Regulating Principle

Ian T. Durham

> *All I did this week was rearrange bits on the internet. I had no real impact on the physical world.*
>
> — Dilbert

Abstract In this essay I develop quantum contextuality as a potential candidate for Wheeler's universal regulating principle, arguing—*contrary* to Wheeler—that this ultimately implies that 'bit' comes from 'it'.

It and Bit

In his Oersted Medal acceptance address in 1983, John Wheeler expressed his view that the primary task of what he called the "coming third era of physics" was the identification of a universal regulating principle arising from a "regularity based on chaos, of 'law without law'" [21]. This third era of physics will clearly require a radical reconsideration of many of our most cherished ideas. Indeed, such reconsiderations have, in recent years, led to the development of categorical quantum mechanics [1, 2], the derivation of physical laws from ordering relations such as posets [5, 14, 15], and the introduction of topos theory to theoretical physics [6, 13], to name but a few. In the latter approach, Döring and Isham have even attempted to answer Heidegger's amorphous question, 'What is a Thing?' [6, 12].

Indeed, what *is* a 'thing' and from whence does it arise? Several approaches to this problem have been proposed. Of particular interest is the aforementioned work of Döring and Isham [6] and that of Wheeler [22] himself. Both question what is seemingly one of the unassailably solid pillars upon which modern science is constructed: real numbers and, most notably, their continuity as *physical* fact. Döring and Isham

I.T. Durham (✉)
Department of Physics, Saint Anselm College, Manchester, NH 03102, USA
e-mail: idurham@anselm.edu

A. Aguirre et al. (eds.), *It From Bit or Bit From It?*,
The Frontiers Collection, DOI 10.1007/978-3-319-12946-4_18

have suggested that physical quantities need not be real-valued [6]. Their argument against the supposition of real-valuedness for physically measurable quantities is partly based on a line of reasoning that connects measuring devices with continuous, smooth manifolds and equates such manifolds with 'classicality'. Wheeler is more blunt; he flatly states "no continuum" [22].

The concept of a discrete physical reality is a very old idea. The early atomists of ancient India and Greece theorized that nature consisted of two fundamental concepts: *atom*, which was the presence of something, and *void* which was the presence of nothing. The former represented physical reality in its most basic form. In the context of modern physics, these concepts have a deep relation to the dual notions of space and time: anything that is physically real *occupies* space and time, while the complete absence of any such occupation corresponds to a vacuum.[1] In order to 'occupy' space and time, a physical 'thing' must generally possess measurable properties or characteristics that provide a means by which that 'occupation' may be measured. As philosopher Eugene Gendlin has noted, Heidegger's notion of 'thing' is really an explanatory approach that "renders whatever we study as some thing in space, located over there, subsisting separate from … us" [11]. In other words, a 'thing' occupies space and time.

Feynman took the position that anything that possesses energy and momentum[2] is physically real, i.e. particles and fields[3] [10]. More generally, Eddington viewed particles and fields as carriers of sets of "variates" [9]. Mathematically, such variates are manifest in symmetries that represent degrees of freedom of the overall state space of a particle or field with each symmetry making up a sub-space. These symmetries, very generally, provide various means by which particles and fields, along with their configurations and interactions, may be distinguished from one another. For the purposes of this essay, I shall refer generally to anything that 'occupies' space and time as *matter-energy*.

Distinguishability, of course, is at the heart of information. As Schumacher and Westmoreland note, "*Information* is the ability to distinguish reliably between possible alternatives" [18]. In this sense, information is encoded in the properties of the particles. Or, as Wheeler saw it, the act of distinguishing one alternative from another actually gives rise to the particles themselves, hence his use of the term "participatory universe". His approach began with the working hypothesis that

> every *it*—every particle, every field of force, even the spacetime continuum itself—derives its function, its meaning, its very existence entirely—even if in some contexts indirectly—from the apparatus-elicited answers to yes-or-no questions, binary choices, *bits* [22].

Döring and Isham formalize this in the notion of a *topos* which is a type of mathematical structure known as a *category*. One way to think of a category is as

---

[1] It is worth noting that in quantum theory the vacuum may be represented by a quantum state $|vac\rangle$. This would seem to blur the distinction between 'being' and 'nothingness,' but we will leave that discussion for another time.

[2] We will not concern ourselves in this essay with the nature of momentum and energy.

[3] Quantum field theory has rendered the difference between particle and field virtually meaningless: a particle *is* the quantization of a field.

a set of objects that has some connective pattern between the objects [20]. Another way to describe a category is as a mathematical structure consisting of objects and arrows [3]. A topos contains two special objects: a *state object* s, and a *quantity-value object* v. A given physical quantity q is represented by an arrow $q : s \to v$ in the topos. As Döring and Isham note, "[w]hatever meaning can be ascribed to the concept of the 'value' of a physical quantity is encoded in (or derived from) this representation" [6]. In this way a 'thing' is then somewhat loosely defined as a bundle of properties wherein these properties refer to values of physical quantities. This is more abstract than Eddington's view as it is not at all clear from this whether it necessarily implies an occupation of space and time. From Wheeler's perspective, these quantities represented the answers to questions we put to nature. At the most fundamental level, he believed that all such questions could be reduced to those for which 'yes' or 'no' were the only possible answers and would thus be represented by a binary digit, i.e. a *bit*.[4] Hence, *it from bit*, according to Wheeler, and thus all physical 'things' are ultimately information-theoretic in origin.

Information Content

A more formal, "rigorously qualitative" definition of information can be given in terms of the order on a *domain* [16]. A domain $(D, \sqsubseteq)$ is a set of objects D together with a partial order $\sqsubseteq$ that includes certain intrinsic notions of completeness and approximation that are defined by this order. For instance, consider two objects $x, y \in D$. The statement $x \sqsubseteq y$ essentially says that x contains (or carries) some (possibly all) information about y., i.e. y is "more informative" than x [16]. For example, a Honus Wagner baseball card contains information about Honus Wagner.[5] Clearly Honus Wagner himself would be far more informative about his life than his baseball card. In the event that x *does* contain the full information about y, then $x = y$ and x is said to be a *maximal element* (object) of the domain, in which case it is an example of an *ideal element*. An object that is not ideal is said to be *partial*. So, given a domain that includes both Honus Wagner and his baseball card, Honus Wagner would be a maximal element while his baseball card would be partial.

We understand a measurement to be a particular type of mapping on a domain that formalizes the concept of *information content*. For a map to meaningfully measure information content, it must, *at a minimum*, be able to distinguish between those elements that it claims are maximally informative (recall the definition of information given by Schumacher and Westmoreland). The details of the formalism are beyond the scope of this essay, but the main point is that the formalism implies the existence of

[4] The first use of the word 'bit' in the sense of a binary digit was in Claude Shannon's seminal 1948 paper on information theory in which he ascribed the origin of the term to John Tukey who had written a memo on which the term 'binary digit' had been contracted to 'bit' [19].

[5] Famously, the T206 Honus Wagner card, distributed between 1909 and 1911, is the most expensive trading card in history, one having sold in 2007 for $2.8 million.

a purely structural relationship between two different classes of informative objects, neither of which need consist of numbers. We may more formally define this as a map $\mu : D \twoheadrightarrow E$ where it is said that μ *reflects* properties of simpler objects E onto more complex objects D [5]. For instance, the set of characteristics that describe Honus Wagner, say as they appear on his baseball card, would be elements of E whereas the *actual* set of characteristics that 'are' Honus Wagner would be elements of D. The act of looking at the baseball card is then a measurement μ and amounts to *inferring* (i.e. 'reflecting') something about Honus Wagner by looking at his baseball card. In a sense, then, μ is just a function that assigns a 'value' to each 'informative object' on a domain that measures its amount of partiality where we understand that 'value' does not necessarily have to mean 'a number' (e.g. it could be the portrait of Honus Wagner that appears on his baseball card) [5, 16]. Given that a physical quantity q really has no meaning outside the act of measurement μ, we can view q as a specific instantiation or representation of μ where $s \in D$ (s is an element of D) and $v \in E$ (v is an element of E). Combining this formalism with Wheeler's assertion that the answers to all *fundamental* measurement 'questions' are binary, a 'bit' is best understood as a quantity-value object v while the numerical result embodied by the bit is $q \in \{0, 1\}$ (i.e. the value of q comes from the set of numbers consisting of 0 and 1).

This is a *crucial* distinction: **a bit is *not* the same thing as its value**. To see why, suppose we were presented with the result of a particular measurement and the numerical value of that result was 1. Further suppose that this is all the information we have about the measurement. We cannot, with certainty, say that 1 is the value of a bit since it could equally well be the value of a 'trit' (that is anything that may exist in one of only *three* mutually exclusive states) or any other '-it', for that matter. In order for us to know that we have been given the value of a *bit*, we must know that the domain of the quantity-value object is $[0, 1] \in \mathbb{N}$ where $\mathbb{N}$ is the set of all natural numbers.[6] Hence, a bit (and any other '-it') is really defined by a *domain*.

Definition 1 (*Bit*) A 'bit' is any instantiation of the domain $[0, 1] \in \mathbb{N}$ in which the values of the domain represent mutually exclusive[7] states.

Information, as defined by Schumacher and Westmoreland, may then be quantified by the probability $P_S(q)$ of successfully determining q.

We refer to knowledge obtained from the act of measurement as *a posteriori* knowledge whereas any *fore*-knowledge of a system prior to an act of measurement is referred to as *a priori* knowledge [9]. For example, prior to opening a pack of baseball trading cards, we fully expect that the pack will only contain *baseball* trading cards as opposed to trading cards for some other sport. Once the pack has been opened and the cards examined, we now know exactly *which* cards are contained within.

---

[6] Again, this notation is meant to formalize the notion that the only values that q may take are 0 and 1.

[7] The requirement of mutual exclusivity is used to distinguish a 'bit' from a 'qubit' where the latter allows for superpositions of 0 and 1.

The knowledge that the pack contains only *baseball* cards is *a priori* whereas the knowledge that the pack, for example, contains a Honus Wagner card is *a posteriori*.

The difference between pre- and post-knowledge, as described here, are due to Eddington [9]. Technically, Eddington referred to *a priori* knowledge as any knowledge that is derived from a study of the actual *procedure* of measurement [9]. In that sense, baseball cards are less illustrative than quantum states. Consider the quantum state

$$|\psi_{a'}\rangle = \sum_{a'} c_{a'}|a'\rangle$$

where $c_{a'}$ is a complex number and $|a'\rangle$ represents a set of basis vectors for some spin axis a'. The state, $|\psi_{a'}\rangle$, represents a state of *a priori* knowledge about an element of the universe since it is not the result of a measurement but rather some observation about the system that indicates the possible measurements that could be made on that system. This is analogous to knowing that a pack of trading cards contains *baseball* cards, but not knowing which specific cards it contains.

Now suppose that we perform a measurement, $\mathbf{S}_z$, on this state such that $|a'\rangle = \mathbf{S}_z|\psi_{a'}\rangle = +\frac{\hbar}{2}|z+\rangle$, i.e. our measurement of the spin along the z-axis yields a value of $+\hbar/2$ with certainty. In this case, the state $|a'\rangle$ represents a state of *a posteriori* knowledge about an element of the universe because it provides information about the *actual* state of the system not just the *possible* state or states of the system.

What is the origin of *a priori* knowledge? How do we know that $|a'\rangle$ represents a spin state as opposed to, say, an energy state or momentum state? Analogously, how do we know that a pack of trading cards specifically contains *baseball* cards? In the latter (and purely classical) case, it is clear that some sort of 'measurement' (in a loose sense) had to have taken place, i.e. the package presumably has some identifying characteristics on it in order to differentiate it from other types of trading cards. Reading the package essentially constitutes an act of measurement. Thus there is really a *sequence* of processes that leads to maximal knowledge of a system. If we start with a complete lack of knowledge such that our sequence of measurements could lead us to literally *any* final result—a Honus Wagner baseball card, a fifty-seven-year-old elephant, or a sixteen-inch diameter pizza—each measurement reduces the range of possibilities from a nearly infinite number down to just one in the end. Thus, every time we make a measurement, we further refine our knowledge of the system, increasing the amount of information we have collected and decreasing the amount of information that we lack.

A *lack* of information about a system is typically quantified via statistical entropy since it is conveniently zero when the state is exactly known. Specifically, given an object $x \in D$ and a measurement $\mu : D \to E$, the Shannon entropy is given as [5, 16]

$$\mu x = -\sum_{i=1}^{n} x_i \log x_i \quad \text{with} \quad x \sqsubseteq y \Rightarrow \mu x \geq \mu y$$

where for some value $n = N$, $x = y$ and $\mu x = \mu y$. So if $\mu : D \rightarrow E$ is a measurement and x is an object that it measures, then the mathematical statement $\mu x \in \max(E) \Rightarrow x \in \max(D)$ says that when μx reaches its maximum value, then we have obtained as much information as we can about x. Thus, if $s \equiv y$ and $v \equiv x$, as our knowledge of the value object v increases, it approaches the maximal element (the state object s) which is mathematically written $v \rightarrow s$. Simultaneously the entropy decreases such that $\mu v \rightarrow \mu s$ and μ is said to be *monotone*. Note that, whereas entropy is a measure of information *content*, probability, in the manner described above, is a measure of information itself. Thus, subject to a few 'moderate' hypotheses, information behaves in the same manner as its content [5].

As an example, consider a jigsaw puzzle that contains a binary message that is only decipherable when the puzzle has been completed where on each puzzle piece is printed the value of exactly one bit (i.e. a 0 or a 1). We may represent the state of the puzzle's message as s. Each time we place a puzzle piece represented as v_i, the partially completed puzzle represents a different value object, v and we gain one new bit, q_n, of information. When we have placed the final piece, $n = N$ in which case we have obtained maximal knowledge of the message and μv will have reached a minimum (in fact it should be zero unless we are missing a piece of the puzzle).

In classical deterministic physics, we typically assume that given a complete set of *a priori* knowledge about a system, all *a posteriori* knowledge about that system may be inferred. In other words, in such a system, while N may (or may not) be infinite, there may a threshold v_{min} such that if $v_{\text{min}} \leq v$, s may be predicted with *near* certainty. For example, it may be that we can accurately predict the puzzle's message with only some fraction of the pieces having been assembled. As another example, suppose that the information about the state of a system is fully encoded in the fraction $\frac{5}{6} = 0.8\bar{3}$. Clearly knowledge of just one significant digit is not enough information to predict the state with anywhere near certainty since, for example, $0.8 = \frac{4}{5}$. While perfect certainty in this example is impossible since $\frac{5}{6}$ is a non-terminating decimal fraction, we can at least establish a limit such that, at some point, we may say with confidence that $v \approx s$ (i.e. at some point we can be fairly certain that the state is $\frac{5}{6}$.

A universe whose future states may be predicted with certainty based on a complete knowledge of its prior states may be said to be *physically deterministic*. We may phrase the condition corresponding to physical determinism in a more rigorous and mathematical manner in terms of state objects and quantity-value objects as follows.

Condition 1 (Physical determinism) Let $u \equiv s$ be the ideal element on the domain of physical measurements that provide information about the universe. Then, if u is static and either N is finite or u is predictable then, $v_{\text{min}} \leq v \Rightarrow v \rightarrow u$.

A hypothetically omniscient being who happens to be in possession of $q_{\text{min}} \leq q_n$ bits, where $q_{\text{min}} : u \rightarrow v_{\text{min}}$ (i.e. possesses enough bits of information about the universe to fully predict its future states), is known as Laplace's demon.

One of the assumptions of physical determinism is that some properties are considered to be immutable—once a Honus Wagner baseball card, always a Honus Wagner baseball card. This is not without its problems as it implies that $\mu v \rightarrow \mu u$, i.e. as we obtain more and more information about the universe, its entropy should *decrease*. The second law of thermodynamics tells us, of course, that the exact opposite is actually happening and as Eddington famously said, "if your theory is found to be against the second law of thermodynamics I can give you no hope; there is nothing for it but to collapse in deepest humiliation" [8]. Something is clearly amiss.

The problem lies in the seemingly innocuous assumption that by increasing our knowledge of a physical system we will necessarily arrive at a full description of the system, i.e. that, as $n \rightarrow N$, it must be that $v \rightarrow s$. For this to be the case, s would have to be determinable. Consider once again the jigsaw puzzle containing a binary message. For s to be determinable, it must be *static* (e.g. the message cannot change as we assemble the puzzle since, if it did, the pieces would need to be reorganized) *and* either N would have to be finite (e.g. the puzzle would have to contain a finite number of pieces) or s would have to be predictable in some manner (e.g. the message would have to have a predictable pattern). While in some cases s may not be determinable with perfect certainty, as I noted previously, in some cases we can establish a limit whereby we may say with some degree of confidence that $v \approx s$. This is frequently the case in classical systems. To use an old adage, "close only counts in horseshoes and hand grenades"—and classical physics. Of course things become a bit more difficult at the quantum level.

Contextuality

Given two objects $x, y \in D$, the statement $x \preceq y$ is read "*x approximates y*".[8] It means that x carries some *essential* information about y where we can think of 'essential' as being synonymous with 'indispensable'. In other words, while x may not carry *all* the information about y, it carries information that is *necessary*. For instance, in the jigsaw puzzle example, there may be certain bits of the message that are indispensable in order for it to be read or comprehended. Any piece that has the value of an indispensable bit printed on it would then be essential.

The statement $x \preceq y$ is context-dependent. Consider three objects, $x, y, z \in D$. Suppose that $x \preceq y$ and that $y \sqsubseteq z$. This means that x carries *essential* information about y and y carries some (not necessarily essential) information about z. In order for us to conclude from this that $x \preceq z$, we would need to know that the statement $y \sqsubseteq z$ is being made in the same *context* as $x \preceq y$ where we take 'context' to mean a 'setting within which we make a statement or measurement'. The results of classical measurements are elements of *continuous domains* which means that classical variables that correspond to measurements may take a continuous (as opposed to discrete) range of values. Approximation (as defined above) on continuous domains is said to be *context independent* [5]. This means that for classical measurements, it is automatically true that if $x \preceq y$ and $y \sqsubseteq z$, then $x \preceq z$. So, for instance, if a certain

[8] The notation $\ll$ is standard but, given the more general audience of this essay, I have adopted $\preceq$ so as to clearly distinguish it from the usual meaning of $\ll$ in inequalities.

piece of our aforementioned jigsaw puzzle is essential, it is essential regardless of how or where (or even when) we assemble the puzzle. Given the limitations on space, I will refer those interested in a more basic explanation of contextuality to Ref. [4].

In quantum systems a 'context' is related to a measurement basis.[9] One way to describe a context, then, is as a domain $(\Omega[\mathsf{m}], \sqsubseteq)$ where for $v_i, s \in \Omega[\mathsf{m}]$ and $v_i \sqsubseteq s$, a specific measurement yields $q_i[\mathsf{m}] : s[\mathsf{m}] \to v_i[\mathsf{m}]$ and m identifies the measurement basis. For an orthonormal n-dimensional basis, we may write $v_1[\mathsf{m}] \perp v_2[\mathsf{m}] \perp \cdots \perp v_n[\mathsf{m}]$ where the symbol $\perp$ is used to indicate the fact that any two v_i in such a basis represent mutually exclusive results. For example, in a two dimensional basis that yields measurement values q_1 and q_2, we have $q_1 : s \to v_1$ and $q_2 : s \to v_2$ where $v_1 \perp v_2$. The values q_1 and q_2 could quite literally be anything as long as it is clear by the device producing the result that they are mutually exclusive. For example, these values might be 0 and 1 or $+1$ and -1 or even Heads and Tails (e.g. for a coin—see Fig. 18.1 for an example involving colors). Thus the quantity-value objects represent the more abstract elements of the basis and are really where the orthogonality manifests itself, i.e. any two mutually exclusive results can be said to be orthogonal. As such, orthogonality, as described here, captures what it means for two 'objects' (elements, states, measurements, etc.) to truly be *distinct*: any two objects that are not orthogonal must share something in common. A more detailed mathematical treatment may be found in Ref. [7].

Now consider a sequence of three devices that measure the spin of a spin-$\frac{1}{2}$ particle along axes a, b, and c. Note that these axes themselves do not necessarily need to be in any way perpendicular to one another. Each represents a *separate* orthonormal basis for which there are only two possible, mutually exclusive results of the measurement. Suppose, then, that the state of the particle exiting the second device is $|b-\rangle$, as shown in Fig. 18.1. Quantum mechanics tells us that the associated probabilities for the results of the third measurement are $P_c(+) = \sin^2 \frac{1}{2}\theta_{bc}$ and $P_c(-) = \cos^2 \frac{1}{2}\theta_{bc}$ where θ_{bc} is the angle between the b and c axes. For example, suppose $\theta_{bc} = 90°$. Then $P_c(+) = P_c(-) = 0.5$ which corresponds to complete randomness (within the basis), i.e. both outcomes are equally probable. This means that we can't obtain any useful information about axis b via this measurement. As is well-known, however,

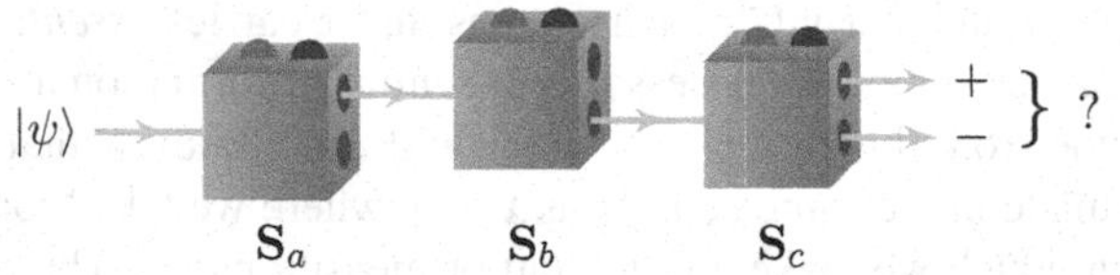

Fig. 18.1 Each box represents a measurement of the spin for a spin-$\frac{1}{2}$ particle along some axis with the top output indicating that the state is *aligned* (+) with the measurement axis, and the bottom output indicating that the state is *anti-aligned* (−) with the measurement axis. *Red* and *blue lights* on the *top* simply indicate to the experimenter which of the two results is obtained (e.g. *red* might indicate aligned and *blue* might indicate anti-aligned)

[9] We point those readers interested in a refresher on measurements and bases in quantum mechanics to Ref. [18].

if the angle is anything *other* than 90°, we *can* make certain deductions about axis b based solely on our measurement along the c axis. Mathematically we can quantify the case in which the angle is 90° by writing $s[b] \cap s[c] = 0$ which means that the state objects for the particle in each instance share nothing in common. Conversely, if $\theta_{bc} = 0°$ which means that $b = c$, then it is as if we are merely confirming the second measurement: we are guaranteed to find that the particle is in the state $|c-\rangle$. This, of course, is the exact opposite of perfect randomness and so we can write $s[b] \cap s[c] = 1$ corresponding to full knowledge of axis b.

In a sense, then, the statement $s[\mathsf{m}] \cap s[\mathsf{n}]$ *quantifies* contextuality and is a bit like a dot product between two vectors with one crucial difference. Suppose that $\theta_{ab} = \theta_{bc} = 90°$ and that $a = c$, i.e. they represent the same axis. Let us suppose that the state subsequent to the first measurement is, as in Fig. 18.1, $|a+\rangle$ and that the state subsequent to the second measurement is, also as in Fig. 18.1, $|b-\rangle$ (in fact it doesn't really matter if it is $|b-\rangle$ or $|b+\rangle$ as long as the angles are as described). We already know that the probabilities for the results of thc third measurement are $P_c(+) = P_c(-) = 0.5$. This means that it is entirely possible for the state subsequent to the *third* measurement to be $|c-\rangle = |a-\rangle$. Clearly $s[a] \cap s[c] = 0$ *even though a and c are the same axis*! Thus contextuality provides a means by which a quantum state can essentially be 'reset'. Alternatively one could say that the particle has no 'memory' of having been in the $|a+\rangle$ state. For a more rigorous technical discussion, see Ref. [7].

Note that while the result of a given measurement may be completely random, the result is always associated with the domain of measurement i.e. we never find a measurement result in a basis *other* than the one in which we choose to make the measurement. For example, if we open a pack of baseball cards we know we won't find that it contains playing cards (or, in even simpler terms, you never find an orange growing on an apple tree[10]).

Quantum states are more complicated than classical states in that they may be represented by a density matrix with complex entries as opposed to as a simple number. The result of a quantum measurement, however, will always produce a result that lies in the domain of *classical* quantity-value objects. For example, though a qubit may exist in some mixture of $|0\rangle$ and $|1\rangle$, when measured it is always found to be in *either* $|0\rangle$ *or* $|1\rangle$. This is because prior to measurement, quantum states may be *informationally isolated* whereas the act of measurement renders them non-isolated which eliminates any superposition or mixed states [18]. To put it another way, the domain of classical states D is *strictly* smaller than the domain of quantum states Ω. I will write this $D \sqsubset \Omega$ where I take $D \sqsubset \Omega$ to mean "D carries some, but strictly *not* all, information about Ω".[11] Thus a quantum measurement is a map $q : \Omega \to D$ and the von Neumann entropy

$$\sigma\rho = -\text{tr}(\rho \log \rho)$$

---

[10] If you ever do, run like hell. The zombies are coming.

[11] This is non-standard notation that I introduce here for the sake of simplifying the presentation.

on quantum states is thus also a measurement $\sigma : \Omega \to E$ where σ factors as $\sigma = \mu \circ q$ and $\mu : D \to E$ is a classical measurement [5]. As the example given in Fig. 18.1 demonstrates, the loss of informational isolation is intimately related to contextuality: prior to making a measurement, the system is informationally isolated and thus can exist in any number of possible states. Once the measurement basis is chosen and the measurement is made, the isolation and thus any uncertainty about the state, is lost. But this means that quantum states can store more information than their classical counterparts. Thus contextuality provides for two important features in quantum states: they may be 'reset' and it is possible for them to store more information than classical states.

Now consider a system for which all measurements are projective. The term 'projective' here is intentionally suggestive. As an analogy, one could think of the process of determining the state of a system via projective measurements to being a bit like trying to determine the shape of a continuously changing object by observing the shapes of the shadows it casts in various planes. For projective measurements, the entropy of the state *after* a measurement, $\sigma\rho'$, is, in fact, *greater* than or equal to the entropy of the state *prior* to the measurement, $\sigma\rho$ with equality only when $\rho' = \rho$, i.e. when the state does not change [17]. It is really *contextuality*, then, that ultimately leads to an increase in entropy in systems such as the one described in Fig. 18.1. If the measurements are ultimately two-dimensional, as in Fig. 18.1, then they are equivalent to the sorts of 'yes/no' questions that Wheeler has suggested form the core of the universe itself [22], suggesting that the second law of thermodynamics is a result of quantum contextuality. According to Condition 1, because u is not static, this further suggests that the universe as a whole must *not* be physically deterministic! For a more rigorous discussion of this assertion, see Ref. [7].

It or Bit?

These considerations then bring us back to Wheeler's original declaration that every 'it' derives its very existence from 'bits' of information. Let us suppose that the answers to these 'yes/no' questions proposed by Wheeler truly are fundamental or nearly fundamental in some manner. Is it 'it from bit' or 'bit from it'? While the entropy—information content—of the universe is constantly increasing (assuming the context is always changing with the questions since, if it didn't, we would live in a rather boring universe), the matter-energy content of the universe is known to be constant. It would seem that if Wheeler were literally correct, this latter point should not be true. In other words, if 'it' truly—literally—comes from 'bit' and the number of bits of information in the universe is always increasing, why doesn't this result in the creation of at least *some* new matter-energy? Even if not *every* new bit of information necessarily led to some new 'it,' it seems reasonable to assume that at least *some* would. The only sensible conclusion, then, is that 'bit' comes from 'it'. Perhaps, then, quantum contextuality serves as an example of Wheeler's 'law without law,' particularly since it may give rise to the second law of thermodynamics [21]:

ordered classical systems emerge from disordered quantum systems. Indeed, perhaps the key to understanding the universe is lying right under our very noses.

References

1. S. Abramsky, B. Coecke, A categorical semantics of quantum protocols, in *Proceedings of the 19th IEEE conference on Logic in Computer Science (LiCS'04)*. IEEE Computer Science Press (2004)
2. S. Abramsky, B. Coecke, Categorical quantum mechanics, in *Handbook of Quantum Logic and Quantum Structures*, vol. II (Elsevier, 2008)
3. S. Awodey, *Category Theory*, 7th edn. (Oxford University Press, Oxford, 2010)
4. D. Bacon, The contextuality of quantum theory in ten minutes (2008). http://scienceblogs.com/pontiff/2008/01/17/contextuality-of-quantum-theor/
5. B. Coecke, K. Martin, A partial order on classical and quantum states. Lect. Notes Phys. **813**, 593–683 (2011)
6. A. Döring, C. Isham, "What is a Thing?": Topos theory in the foundations of physics. Lect. Notes Phys. **813**, 753–937 (2011)
7. I.T. Durham, An order-theoretic quantification of contextuality. Information **5**(3), 508–525 (2014)
8. A.S. Eddington, *The Nature of the Physical World* (Cambridge University Press, Cambridge, 1928)
9. A.S. Eddington, *The Philosophy of Physical Science* (Cambridge University Press, Cambridge, 1939)
10. R.P. Feynman, R.B. Leighton, M. Sands, *Feynman's Lectures on Physics*, vol. 1 (Addison Wesley, Reading, 1963)
11. E.T. Gendlin, What is a thing?, in *An Analysis of Martin Heidegger's What is a Thing?*, ed. by M. Heidegger (Henry Regnery, Chicago, 1967), pp. 247–296
12. M. Heidegger, *What is a Thing?* (Regenery/Gateway, Indiana, 1967)
13. C. Isham, Topos methods in the foundations of physics, in *Deep Beauty*, ed. by H. Halvorsen (Cambridge University Press, Cambridge, 2010)
14. K.H. Knuth, Deriving laws from ordering relations, in *Bayesian Inference and Maximum Entropy Methods in Science and Engineering (AIP Conference Proceedings)*, ed. by Y. Zhai, G.J. Erickson (American Institute of Physics, Melville, 2003)
15. K.H. Knuth, N. Bahrenyi. A derivation of special relativity from causal sets (2010). arXiv:1005.4172
16. K. Martin, Domain theory and measurement. Lect. Notes Phys. **813**, 491–591 (2011)
17. M.A. Nielsen, I.L. Chuang, *Quantum Computation and Quantum Information* (Cambridge University Press, Cambridge, 2000)
18. B. Schumacher, M. Westmoreland, *Quantum Processes, Systems, and Information* (Cambridge University Press, Cambridge, 2010)
19. C.E. Shannon, A mathematical theory of communication. Bell Syst. Tech. J. **27**(3), 379–423 (1948)
20. D.I. Spivak, Category theory for scientists (2013). arXiv:1302.6946
21. J.A. Wheeler, On recognizing 'law without law,' oersted medal response at the joint aps-aapt meeting, New York, 25 January 1983. Am. J. Phys. **51**(5), 398 (1983)
22. J.A. Wheeler, Information, physics, quantum: the search for links, in *Complexity, Entropy and the Physics of Information*, vol. VIII, Santa Fe Institute Studies in the Sciences of Complexity, ed. by W.H. Zurek (Addison Wesley, Redwood City, 1990), pp. 3–28

Chapter 19
It from Bit from It from Bit … Nature and Nonlinear Logic

Wm. C. McHarris

Abstract For the last decade I have been demonstrating that many of the so-called paradoxes generated by the Copenhagen interpretation of quantum mechanics have less puzzling analogs in nonlinear dynamics and chaos theory. This raises questions about the possibility of nonlinearities inherent in the foundations of quantum theory. Since most of us do not think intuitively with nonlinear logic, I take this opportunity to dwell on several peculiarities of nonlinear dynamics and chaos: nonlinear logic itself and the possible connection of infinite nonlinear regressions with free will. Superficially, nonlinear dynamics can be every bit as counterintuitive as quantum theory; yet, its apparent paradoxes are more amenable to logical analysis. As a result, using nonlinear dynamics to resolve quantum paradoxes winds up being more straightforward than many of the alternative interpretations currently being formulated to replace the orthodox interpretation. Chaos theory could be a candidate for bridging the gap between the determinism so dear to Einstein and the statistical interpretation of the Copenhagen School: For deterministic chaos is indeed deterministic; however, intrinsic limitations on precision in measuring initial conditions necessitate analyzing it statistically. Einstein and Bohr both could have been correct in their debates.

> Quantum mechanics is the very epitome of a linear science.
>
> Before observation a system can exist in a superposition of quantum states, such as Schrödinger's cat's being both dead and alive.
>
> A particle such as an electron is intrinsically neither particle nor wave but exists in an epistemological limbo: if one performs an experiment to measure particle-like properties, it cooperates and acts as a particle; if one seeks wave-like properties, it behaves as a wave.
>
> Classical mechanics is a meso- or macro-scale approximation to quantum mechanics, with the Correspondence Principle connecting the two.
>
> The experimental violations of Bell-type inequalities demonstrate that local reality fails and entangled particles can communicate faster than the speed of light—contrary to the laws of relativity. Einstein's 'spooky action at a distance' does in fact occur.

W.C. McHarris (✉)
Departments of Chemistry and Physics/Astronomy,
Michigan State University, East Lansing, MI 48824, USA
e-mail: mcharris@chemistry.msu.edu

A. Aguirre et al. (eds.), *It From Bit or Bit From It?*,
The Frontiers Collection, DOI 10.1007/978-3-319-12946-4_19

For many decades statements such as these formed something akin to a physicist's bible. Despite the many paradoxes that result from this orthodox Copenhagen interpretation, few dared to doubt its validity. The holdouts, such as Einstein, Schrödinger, deBroglie, and later Bohm et al., were judged to be past their prime—after all, Bohr had vanquished Einstein in their debates and in his response to the EPR paper of 1935 [1]. Reductionism reigned supreme!

Then along came quantum information science with the enticement of super-fast quantum computing. Interest revived in the fundamentals of quantum mechanics—now and then people began to question tenets of the Copenhagen interpretation and to proffer alternatives. There were even a few attempts to deal with possible nonlinearities intrinsic to quantum mechanics, but these were treated as perturbations on a basically linear theory [2–4], a situation in which the extreme peculiarities of nonlinear dynamics, demonstrated by systems in or at the edge of chaos, cannot come to pass [5]. Strongly nonlinear systems, in which such characteristics are allowed to develop, can seem most peculiar—indeed, they can appear to be every bit as counterintuitive and paradoxical as quantum mechanics itself. However, just as with the oddities of relativity, upon more detailed examination these nonlinear peculiarities can be resolved in a considerably more logical manner than can the paradoxes of orthodox quantum mechanics.

During the past decade I have published a number of papers demonstrating that many of the so-called paradoxes generated by the Copenhagen interpretation of quantum mechanics have parallels in nonlinear dynamics and chaos ([6, 7], and references therein). For example, the inherent randomness of radioactive decay can be reconciled with observed first-order exponential decay laws through the extreme (exponential) dependence on initial conditions—the so-called "Butterfly Effect". Or, chaotic scattering of classical particles can produce diffraction-like patterns such as those characteristic of waves [8, 9]. And strongly nonlinear classical systems can exhibit correlations—or entanglements!—analogous to those of quantum systems, thereby violating Bell-type inequalities [6, 10]; similarly, nonergodic behavior can easily ape "action at a distance". Indeed, macroscopic nonlinear systems can produce quantization intrinsically, for they often organize themselves into "quantized" or resonant levels, which are governed by eigenvalue equations analogous to the Schrödinger equation.

These parallels and analogies by no means prove that quantum mechanics is fundamentally nonlinear, and this is not the forum in which to delve into their detailed justification. Nevertheless, they do raise important, fundamental questions about the possibilities of strongly nonlinear effects at the heart of quantum mechanics. Strongly nonlinear and/or chaotic systems can be deterministic, i.e., a given set of initial conditions leads deterministically to a specific final state—cause and effect—but because it is normally impossible to determine the initial conditions with sufficient precision to achieve such specific predictions, these systems must be treated statistically. Deterministic chaos provides the determinism so dear to Einstein, yet it must be treated statistically à la Bohr and the Copenhagen School. Perhaps it can provide the bridge between the two viewpoints—in hindsight, both Einstein and Bohr could have been correct in their debates!

During the past fifteen or so years I have given close to fifty lectures on some variation of "Chaos and the Quantum" at universities, national laboratories, and regional-national-international conferences—some to physicists studying the fundamental basis of quantum mechanics, others to sundry scientists studying nonlinear dynamics, chaos, and complexity. I have found very little communication between the two groups: Nonlinear dynamicists know very little about the fundamentals of quantum mechanics, and while physicists may have heard of nonlinear dynamics and chaos, they mostly consider them to lie in an obscure corner of science, with little relevance to their primary goals. At times I have felt as if I had one foot resting on a quantum ice-floe, the other on a chaotic ice-floe—while the two are rapidly drifting apart!

A favorite quip of nonlinear dynamicists is, "Calling most of dynamics *nonlinear dynamics* is like calling all of zoology not concerned with elephants *nonpachydermology*!" Most of nature is indeed nonlinear, filled with various forms of feedback. Nonlinear dynamics and chaos, although a new science, has rapidly gained enormous success in fields as diverse as biology, chemistry, economics, even traffic patterns. The glaring exception has been quantum mechanics, which should make us suspicious. I feel that nonlinear dynamics and chaos theory are changing not only the way we do science, but also the way we view the world (and the universe). Yet most of us have difficulty in thinking nonlinearly or dealing with nonintuitive feedback, just as we have difficulty in coping with concepts such as correlated, Bayesian statistics or with infinite limits. Since this is primarily a physics forum, I limit myself to two straightforward, intentionally over-simplified demonstrations of nonlinear dynamics, illustrating *nonlinear* logic.

Nonlinear Analysis

We are simply not used to thinking in terms of nonlinear logic. Sure, we can deal with simple systems. Take, for example, the deer population in Northern Michigan. A typical environment can only support so many deer. As the population increases, the deer overgraze, causing the forage to decrease, which eventually leads to less healthy, then fewer deer. With fewer deer, the vegetation recovers, which leads again to more deer. A straightforward example of feedback. We would not be surprised to find a herd oscillating between, say, fifty and twenty deer in alternate years. And, with climate change making Northern Michigan more verdant, we could easily understand a trend that saw the same herd increase and oscillate between larger numbers, such as eighty and thirty-two deer. It would merely be a case of different environmental parameters.

But suppose we performed a carefully controlled study and discovered that a given herd oscillated over a longer timespan among four distinct numbers, or even eight numbers—or that the number of deer became completely unpredictable even though we were working under carefully controlled conditions? I wager that such an analysis would discourage most of even the hardiest researchers. This, however, is a relatively simple case of biological feedback, where nonlinear dynamics, then chaos sets in. And we shall see in the next section that it can be explained by a surprisingly *simple* model. *Chaos is the situation where complex, even seemingly inexplicable behavior results from* ***simple*** *systems—and without intervention from external complicating influences.*

Another straightforward example that confounds most people is the mixing—or unmixing—of powders in a nonlinear tumbler, a rotating cylinder having part of one arc replaced by a flattened side [11]. A mixture of salts often separates into distinct bands in such a tumbler. Surely this is a spontaneous decrease in entropy, contrary to everything we have learned about the Second Law of Thermodynamics?! Yet it happens. An even simpler demonstration (gleaned from the same article) is the following: Fill a cylindrical bottle, about 5 cm in diameter × 15–20 cm long, roughly 60 % full of common NaCl. Then add a plastic map tack and an iron hexagonal nut. When you shake the bottle horizontally one of the two comes to the top, but vertically, the other. I leave it to you to determine which is which. A simple-minded demonstration, but one that makes a strong impression on college chemistry students.

A less pleasant onset of chaos can occur when a complicated system has its end goals altered while it is under construction. This has occurred on at least two occasions with major computer programs. Both the U.S. Red Cross Blood Bank software operating system and one of the U.S. Naval Defense systems suffered this fate, when succeeding administrations changed the ultimate goals of the software numerous times. Chaos set in, and the systems became unmanageable and unusable and had to be replaced almost from scratch, at considerable financial loss. (Although not enough data have been made public to reach definite conclusions, it seems likely that the initial collapse of the healthcare.gov website could have resulted partly from similar causes, although many other factors, such as lack of communication between agencies undoubtedly also played major roles.)

Perhaps the most dramatic illustrations of how little we comprehend nonlinear logic arise in computer science from evolutionary computer programs. Initiated at MIT and Stanford, this sort of research is now being performed at many major universities, including MSU. Here is a seemingly straightforward example:

A popular yardstick among computer scientists is to create an efficient program that will sort, say, a set of 100 random numbers into ascending order. To simulate evolution in creating such a program, one first uses a pseudorandom number generator to generate programs consisting of (almost) random sequences of numbers. One can then either use these programs "raw", or to speed up the process, one can retain only instructions at least marginally useful for sorting, such as comparison and exchange instructions. Thus, one begins with a population consisting of, say, 10,000 random programs, each consisting of several hundred instructions.

One then runs and tests these randomly generated programs, which, as expected, do a very poor job at first. Only the "most fit", meaning any programs that show the slightest inclination for sorting, are retained. These are used to create the next generation. This can be done in two ways: First, by inserting random minor variations, corresponding to asexual mutations; second, by "mating" parent programs to create a child program, i.e., by splicing parts of programs together, hoping that useful instructions from each parent occasionally will be inherited and become concentrated. This process can be repeated thousands upon thousands of time with fast parallel processors, and eventually very efficient programs can result. (It should be mentioned in passing that this sort of research is carried out on isolated computers using nonstandard operating systems, for such programs could become dangerous viruses on the internet.)

Some of the results were startling. Danny Hillis, who originated many aspects of these programs, sums it up vividly in his book, *The Pattern on the Stone* [12]:

> I have used simulated evolution to evolve a program to solve specific sorting problems, so I know that the process works as described. In my experiments, I also favored the programs that sorted the test sequences quickly, so that faster programs were more likely to survive. This evolutionary process created very fast sorting programs. For the problems I was interested in, the programs that evolved were actually slightly faster than any of the algorithms described... [standard algorithms]—and, in fact, they were faster at sorting numbers than any program I could have written myself.
>
> **One of the interesting things about the sorting programs that evolved in my experiment is that I do not understand how they work.** [my emphasis] I have carefully examined their instruction sequences, but I do not understand them; I have no simpler explanation of how the programs work than the instruction sequences themselves. It may be that the programs are not understandable—that there is no way to break the operation of the program into a hierarchy of understandable parts. If this is true—if evolution can produce something as simple as a sorting program which is fundamentally incomprehensible—it does not bode well for our prospects of ever understanding the human brain.

This evolutionary process has also been used to design electronic circuits [13], with similar results. Circuits can result that are considerably more efficient than those produced by professional designers. Sometimes they also contain superfluous appendages, apparently leftovers from the evolutionary process. But again, they cannot be analyzed in any straightforward manner. They just work. It should also be mentioned that in all of these processes there is a tendency to reach only local maxima. The introduction of "predator" programs overcomes this, forcing the attainment of global maxima—completely analogous to Darwinian evolution.

In other words, there are logical processes that cannot be understood, much less be broken down into reductionistic, simply analyzable parts. On the positive side, nature is far more intricate and beautiful than we could imagine, using our simple(-minded) models. On the negative side, nonlinear systems tend not to be generalizable (despite Feigenbaum's "Universality" in chaotic systems). Thus, producing, say, a Boolean system for analyzing nonlinear logic may be beyond us.

Infinite Regression and Free Will

Extreme, exponential sensitivity to initial conditions in deterministic chaotic systems means just that—although a given set of initial conditions most certainly leads to a definite final state at a specified future time, in practice it is impossible, even unthinkable, to determine a set of initial conditions with anywhere nearly enough precision

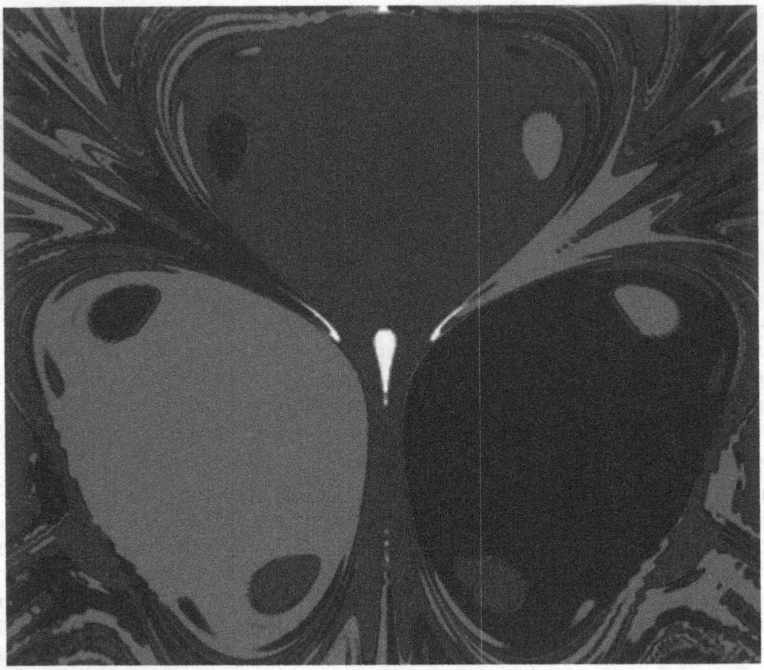

Fig. 19.1 Basins of attraction for a pendulum suspended above three magnets arranged in an equilateral triangle with the "*red*" magnet situated near the top of the *red* basin and "*blue*" and "*green*" magnets in corresponding positions on their basins. It can be proven that between every two colors lies a basin of the third color, ad infinitum. Thus, there are regions where it is physically impossible, no matter how precise the experiment, to predict (other than statistically) where the pendulum will stop. This is an example of the Butterfly Effect

in order to produce a predetermined final state. (In practice, chaoticists get around this presumed limit to the usefulness of chaos theory because other chaotic orbits in phase space "shadow" the desired orbit, making statistical predictions possible. But this is beyond the scope of this essay.) A simple example illustrates this:

Consider a simple three-dimensional iron-bob pendulum free to oscillate above a surface that contains three magnets arranged in a triangle. We color the "basins of attraction" of these three magnets red, blue, and green, as illustrated in Fig. 19.1. That is, if the pendulum is released initially somewhere above the large red blob, it will eventually wind up hovering near the "red" magnet. (Although in this instance the magnets really do attract the bob, the term "basin of attraction" refers to any region in phase space that "surrounds" a point where an orbit is likely to wind up, whether or not the attraction is physical.) There is similar behavior for the "blue" and "green" magnets. But what if it is released near the boundary of the red and green basins—it very likely will wind up hovering above the blue magnet. However, things can get complicated very quickly. It has been shown that between every two colors there is a basin of the third color, ad infinitum. In other words, there are regions where it is physically impossible, no matter how great the precision of one's experiment, to predict where the pendulum will come to rest. This is true whether the pendulum is a toy or a magnificent Foucault pendulum suspended from a many-storied dome. It just takes longer to reach the limit of precision with the Foucault pendulum—we say it takes longer for the uncertainty to catch up with and overwhelm the precision of the measurement. Intricately intwined basins of attraction such as these are known as "riddled basins", and they are the norm rather than the exception. (This is very similar to Newton's method for extracting the complex roots of a third-order polynomial, where there are regions of initial guesses that lead to unexpected roots. The basins of attractions look quite similar, an example of so-called Universality in chaos, where identical or very similar equations can be used to describe quite disparate situations.)

Although the above example is quite simple to visualize, its math is not the simplest, so, in keeping with Occam's Razor, here is the very simplest, most studied example of infinite regression, the **logistic map**. Its history in describing population dynamics goes all the way back to the Belgian mathematician, Verhulst, in the 1840s.

The very simplest model for predicting the next generation population would be to multiply this generation's population x_n by the birth rate A. However, if the birth rate is greater than 1, this leads to Malthusian, exponential growth, clearly unrealistic. One then notes that a given environment can only support so many individuals, so includes the difference between this maximum and the present population as the first, simplest correction. The next generation is thus predicted to be

$$x_{n+1} = Ax_n(1 - x_n).$$

Here x lies in the range of 0–1, where 0 indicates vanishing population (extinction) and 1 is the maximum population. What could be simpler and more predictable? However, as can be seen in Fig. 19.2, this equation can produce some rather unexpected behavior.

First of all, for $A < 1$, the population vanishes, as expected, for all initial values. (We say that here 0 is an attractor of the system for such values of A.) For $A > 1$, after sufficient iterations the population settles down to a single value that monotonically increases as A increases, again as expected. For values of A infinitesimally greater than 3 the final value bifurcates, i.e., alternates between two values. As A continues to increase, further bifurcations occur, with periods of 4, 8, 16, ... (The periods are not simply powers of 2 but incur all the natural numbers in an elaborate number-theoretical sequence called the Sarkovskii ordering.) Finally, at $A > 3.44948\ldots$ chaotic behavior ensues, in which the map never settles down but continues hitting seemingly random values ad infinitum. The "dust" in Fig. 19.2 results not from poor graphics nor from not carrying out sufficient iterations to fill the gaps—it results from infinitely complex detail in the diagram. As can be seen from the successive magnifications, this diagram is said to be "self-similar" (or more correctly, "self-affine", because the frames are not exact copies). In the limit of infinite iterations, the magnification—no matter how great, even infinite—of any portion of the diagram resembles the diagram as a whole.

Perhaps the most dramatic illustrations of infinite regression come with the Mandelbrot set, sometimes billed as the "simplest, most complex object in the universe"!

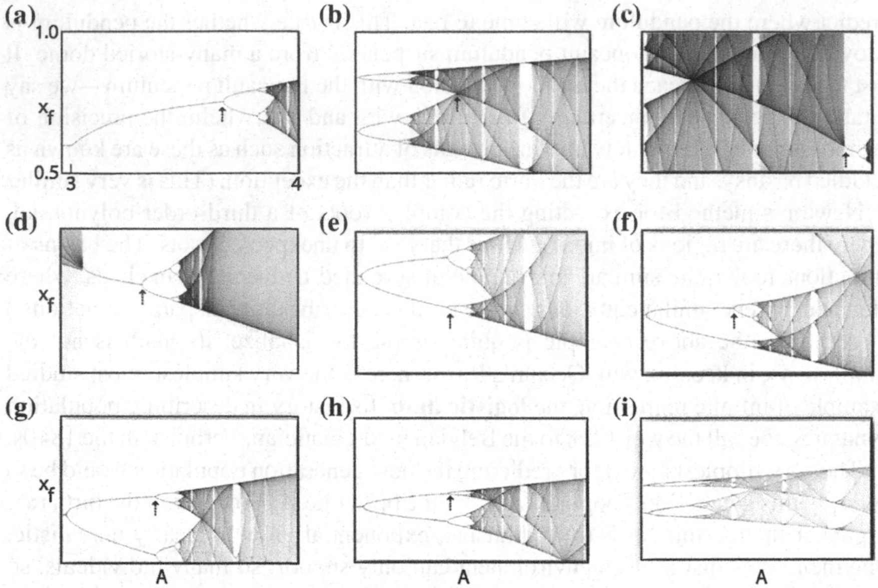

Fig. 19.2 Magnifications showing the self-affinity of the bifurcation diagram for the logistic equation. Here the final population x_f (after an "infinite" number of iterations) is plotted against A, the control parameter or "birthrate". Only the regions of interest are shown—for $A < 1$ the population goes to 0, and for $1 < A > 3$ there is a monotonic increase as A increases. The first panel shows the final population for values of A ranging from just below 3 to 4 (just above $A = 4$ the model breaks down). Successive panels show a portion of each preceding panel near the *arrow* magnified by a factor of roughly 5. Thus, the final panel has been magnified by a factor close to one million

It results from successive mappings of another very simple equation similar to the logistic equation, only this time dealing with the regions of divergence vs convergence for values of the control parameter in the complex plane. There are dozens, if not hundreds of videos available on the internet illustrating this. At present the most extreme and dramatic one takes you through all the intricacies of the Mandelbrot set up to a magnificent magnification of 2^{316} [14]!

Now, if this chaotic behavior were truly random, there would be little point in following through with chaos theory. However, there is a definite, albeit subtle order in chaos. One of the clearest manifestations of this can be seen as the gaps in Fig. 19.2; one that stands out is the large period-3 gap in the vicinity of $A = 3.82$. There are an infinite number of such gaps, persisting all the way down to infinite magnification. This means there are regions of order (periodic behavior) intimately mixed in with the chaotic regions. (To revert to quantum mechanics, although I promised I wouldn't, this could be analogous to duality, where periodic regions represent particle-like behavior, chaotic regions, wave-like behavior.)

What does all of this mean in terms of free will? Simply this: Just as chaos theory can provide a bridge between determinism and statistical behavior, so can it provide a bridge between predestination and free will. Maps having self-affine infinite regression are ubiquitous in the universe, and these maps can be strictly deterministic—a given initial point inevitably leads to a definite final point. Nevertheless, because it is *physically impossible* to determine this initial point with the necessary *infinite precision*, one cannot work backward from the "predestined" final point to its defined cause. Mathematics can state things with certainty—physics cannot. In physics we are fond of stating, "In principle, it can be shown (even proven) that...". In fact, indeed *in principle*, chaos theory shows this not to be true. Principle and practice represent two antipodal worlds—perhaps never the twain shall meet.

Free will has received considerable attention in the fundamentals of quantum mechanics world in the last few years, in particular with the Free Will Theorem of Conway and Kochen [15], which interjects the idea of free will into the behavior of fundamental particles (at the Planck scale). It has been used to castigate certain contemporary interpretations of quantum theory. Could this not be necessary? Could free will be represented by a simpler (remember Occam's Razor) nonlinear infinite regression?

Afterword

It from Bit or Bit from It? Nonlinear dynamics and chaos theory shows us that disparate parts of nature are intimately linked together much more tightly than we could previously have imagined. Wherever there is feedback there is interaction between/among the various component parts. We could well be fooling ourselves with our "straightforward" linear, reductionist models. Could it be significant that chaos theory has had successes in almost every scientific field other than quantum mechanics—to the extent that for decades various respectable scientists have doubted the very existence of true quantum chaos [16]. Could it be that quantum mechanics already contains essential nonlinear elements, and that when we try to apply chaos theory to it, we are going around in a loop? An esthetic advantage of injecting nonlinear dynamics into quantum mechanics is that it is relatively simple and straightforward (Occam's Razor, for the last time), does not lead us into the quagmire of assuming hidden variables (!), and it makes the transition from quantum to classical mechanics smoother, eliminating part of the thorny problem of determining the border between observer and observed. Nonetheless, despite avowed pronouncements from a myriad of prominent and respected physicists, beauty in equations does not make a theory true—or relevant. Only experimental investigation—and the ability of a theory to be falsifiable can do that, which is why nowadays we are seeing a significant number of critical, almost damning expositions such as *The Trouble with Physics* and *Not Even Wrong* [17, 18].

As for information science: "It from Bit or Bit from It?" is a little bit like the problem of the chicken and the egg. One can easily become trapped in an infinite loop. Which brings us back to where we started. It from Bit from It... involves nonlinearity and feedback, which leads us to nonlinear dynamics and chaos, which leads to infinite regressions, which leads to..., ..., ...

References

1. A. Einstein, B. Podolsky, N. Rosen, Phys. Rev. **44**, 777 (1935)
2. S. Weinberg, Ann. Phys. (N. Y.), 194, 336 (1989).
3. N. Gisin, Phys. Lett. **A143**, 1 (1990)
4. B. Mielnik, Phys. Lett. **A289**, 1 (2001)
5. M.J. Feigenbaum, Foreword, in Chaos and Fractals: New Frontiers of Science, 2nd edn., ed. by H.-O. Peitgen, H. Jürgens, D. Saupe (Springer, New York, 2003), p. 1.
6. Wm.C. McHarris, J. Phys.: Conf. Ser. 306, 012050 (2011).
7. Wm.C. McHarris, Complexity, 12(4), 12 (2007).
8. S. Bleher, C. Grebogi, E. Ott, Phys. D **46**, 87 (1990)
9. J.V. José, E.J. Saletan, Classical Dynamics: A Contemporary Approach, Chap. 4, Sect. 4.1.3 (Cambridge University Press, Cambridge, 2002).
10. M. Gell-Mann, C.J. Tsallis (eds.), *Nonextensive Entropy: Interdisciplinary Applications* (Oxford University Press, New York, 2004)
11. T. Shinbrot, F.J. Muzzio, Phys. Today **53**(3), 25 (2000)
12. W.D. Hillis, *The Pattern on the Stone: The Simple Ideas That Make Computers Work* (Basic Books, New York, 1998)
13. J.R. Koza, M.A. Keane, M.J. Streeter, Sci. Am. Feb. 52 (2003).
14. C. Korda, Fractice Mandelbrot Deep Zoom to 2^316 (Bigger than the Universe), http://vimeo.com/6035941
15. J. Conway, S. Kochen, Found. Phys. 36(10), 1441 (2006). A series of six informative lectures by John Conway can be found at http://web.math.princeton.edu/facultypapers/Conway
16. J. Ford, in The New Physics, ed. by P. Davies, Chap. 12 (Cambridge University Press, Cambridge, 1989).
17. L. Smolin, *The Trouble with Physics: The Rise of String Theory, the Fall of a Science, and What Comes Next* (Houghton Mifflin Harcourt, Boston, 2006)
18. P. Woit, *Not Even Wrong: The Failure of String Theory & the Continuing Challenge to Unify the Laws of Physics* (Basic Books, New York, 2006)

Author Biography

Wm. C. McHarris is Professor Emeritus of Chemistry and Physics/Astronomy at Michigan State University. He received his B.A. in chemistry from Oberlin College, then his Ph.D. in nuclear chemistry from the University of California, Berkeley, in the turbulent 1960's. He came directly from graduate school to MSU as Assistant Professor, becoming full Professor at age 32. For most of his career he performed research in nuclear physics/chemistry at the National Superconducting Cyclotron Laboratory, but for the last decade or so has been striving to reconcile chaos theory with quantum mechanics. He is also an organist, carillionneur, ragtime pianist, and published composer.

Appendix
List of Winners

First Prize

Matthew Leifer: *"It From Bit" and the Quantum Probability Rule*[1]

Second Prizes

Angelo Bassi, Saikat Ghosh, & Tejinder Singh: *Information and the Foundations of Quantum Theory*
Carlo Rovelli: *Relative Information at the Foundation of Physics*

Third Prizes

Howard Barnum[2]: *Informational Characterizations of Quantum Theory as Clues to Wheelerian Emergence*
Giacomo D'Ariano: *It From Qubit*
Craig Hogan: *Now Broadcasting in Planck Definition*
Kevin Knuth: *Information-Based Physics and the Influence Network*
Ken Wharton: *Reality, No Matter How You Slice It*

---

[1] From the Foundational Questions Institute website: http://www.fqxi.org/community/essay/winners/2013.1

[2] Howard Barnum's essay is not included in this volume.

A. Aguirre et al. (eds.), *It From Bit or Bit From It?*,
The Frontiers Collection, DOI 10.1007/978-3-319-12946-4

Fourth Prizes

Torsten Asselmeyer-Maluga: *Spacetime Weave—Bit as the Connection Between Its or the Informational Content of Spacetime*
Paul Borrill: *An Insight Into Information, Entanglement and Time*
Ian Durham: *Contextuality: Wheeler's Universal Regulating Principle*
Mark Feeley: *Without Cause*
Sean Gryb: *Is Spacetime Countable?*
William McHarris: *It From Bit From It From Bit... Nature and Nonlinear Logic*
Michel Planat: *It From Qubit: How to Draw Quantum Contextuality*
Yutaka Shikano: *These From Bits*
Douglas Singleton, Elias Vagenas, and Tao Zhu: *Self-Similarity, Conservation of Entropy/Bits and the Black Hole Information Puzzle*
Cristinel Stoica: *The Tao of It and Bit*

Other Honours

Julian Barbour's essay, *Bit from It*, which is included in this volume, placed fourth in the 2011 FQXi Essay Contest, "Is Reality Digital or Analog?"

Titles in this Series

Quantum Mechanics and Gravity
By Mendel Sachs

Quantum-Classical Correspondence
Dynamical Quantization and the Classical Limit
By Dr. A. O. Bolivar

Knowledge and the World: Challenges Beyond the Science Wars
Ed. by M. Carrier, J. Roggenhofer, G. Küppers and P. Blanchard

Quantum-Classical Analogies
By Daniela Dragoman and Mircea Dragoman

Life — As a Matter of Fat
The Emerging Science of Lipidomics
By Ole G. Mouritsen

Quo Vadis Quantum Mechanics?
Ed. by Avshalom C. Elitzur, Shahar Dolev and Nancy Kolenda

Information and Its Role in Nature
By Juan G. Roederer

Extreme Events in Nature and Society
Ed. by Sergio Albeverio, Volker Jentsch and Holger Kantz

The Thermodynamic Machinery of Life
By Michal Kurzynski

Weak Links
The Universal Key to the Stability of Networks and Complex Systems
By Csermely Peter

The Emerging Physics of Consciousness
Ed. by Jack A. Tuszynski

A. Aguirre et al. (eds.), *It From Bit or Bit From It?*,
The Frontiers Collection, DOI 10.1007/978-3-319-12946-4

Quantum Mechanics at the Crossroads
New Perspectives from History, Philosophy and Physics
Ed. by James Evans and Alan S. Thorndike

Mind, Matter and the Implicate Order
By Paavo T.I. Pylkkänen

Particle Metaphysics
A Critical Account of Subatomic Reality
By Brigitte Falkenburg

The Physical Basis of The Direction of Time
By H. Dieter Zeh

Asymmetry: The Foundation of Information
By Scott J. Muller

Decoherence and the Quantum-To-Classical Transition
By Maximilian A. Schlosshauer

The Nonlinear Universe
Chaos, Emergence, Life
By Alwyn C. Scott

Quantum Superposition
Counterintuitive Consequences of Coherence, Entanglement, and Interference
By Mark P. Silverman

Symmetry Rules
How Science and Nature Are Founded on Symmetry
By Joseph Rosen

Mind, Matter and Quantum Mechanics
By Henry P. Stapp

Entanglement, Information, and the Interpretation of Quantum Mechanics
By Gregg Jaeger

Relativity and the Nature of Spacetime
By Vesselin Petkov

The Biological Evolution of Religious Mind and Behavior
Ed. by Eckart Voland and Wulf Schiefenhövel

Homo Novus – A Human Without Illusions
Ed. by Ulrich J. Frey, Charlotte Störmer and Kai P. Willfiihr

Brain-Computer Interfaces
Revolutionizing Human-Computer Interaction
Ed. by Bernhard Graimann, Brendan Allison and Gert Pfurtscheller

The Language Phenomenon
Human Communication from Milliseconds to Millennia
Ed. by P.-M. Binder and K. Smith

The Dual Nature of Life
By Gennadiy Zhegunov

Natural Fabrications
By William Seager

Ultimate Horizons
By Helmut Satz

Physics, Nature and Society
By Joaquín Marro

Extraterrestrial Altruism
Ed. by Douglas A. Vakoch

The Beginning and the End
By Clément Vidal

A Brief History of String Theory
By Dean Rickles

Singularity Hypotheses
Ed. by Amnon H. Eden, James H. Moor, Johnny H. Søraker and Eric Steinhart

Why More Is Different
Philosophical Issues in Condensed Matter Physics and Complex Systems
Ed. by Brigitte Falkenburg and Margaret Morrison

Questioning the Foundations of Physics
Which of Our Fundamental Assumptions Are Wrong?
Ed. by Anthony Aguirre, Brendan Foster and Zeeya Merali

It From Bit or Bit From It?
On Physics and Information
Ed. by Anthony Aguirre, Brendan Foster and Zeeya Meral

Printed in the United States
By Bookmasters